CESE-2019: Applications of Membranes for Sustainability

CESE-2019: Applications of Membranes for Sustainability

Editors

Veeriah Jegatheesan
Chettiyappan Visvanathan
Li Shu
Faisal I. Hai
Ludovic F. Dumée

MDPI • Basel • Beijing • Wuhan • Barcelona • Belgrade • Manchester • Tokyo • Cluj • Tianjin

Editors

Veeriah Jegatheesan
School of Engineering and Water: Effective Technologies and Tools (WETT) Reseach Centre, RMIT University
Australia

Chettiyappan Visvanathan
School of Environment, Resources and Development, Asian Institute of Technology
Thailand

Li Shu
School of Engineering, Edith Cowan University and LJS Environment
Australia

Faisal I. Hai
Stategic Water Infrastructure Lab, School of Civil, Mining and Environmental Engineering, University of Wollongong
Australia

Ludovic F. Dumée
Department of Chemical Engineering, Khalifa University Abu Dhabi
UAE

Editorial Office
MDPI
St. Alban-Anlage 66
4052 Basel, Switzerland

This is a reprint of articles from the Special Issue published online in the open access journal *Membranes* (ISSN 2077-0375) (available at: https://www.mdpi.com/journal/membranes/special_issues/CESE_2019).

For citation purposes, cite each article independently as indicated on the article page online and as indicated below:

LastName, A.A.; LastName, B.B.; LastName, C.C. Article Title. *Journal Name* **Year**, *Volume Number*, Page Range.

ISBN 978-3-0365-3823-5 (Hbk)
ISBN 978-3-0365-3824-2 (PDF)

Cover image courtesy of Jega Veeriah Jegatheesan.

© 2022 by the authors. Articles in this book are Open Access and distributed under the Creative Commons Attribution (CC BY) license, which allows users to download, copy and build upon published articles, as long as the author and publisher are properly credited, which ensures maximum dissemination and a wider impact of our publications.

The book as a whole is distributed by MDPI under the terms and conditions of the Creative Commons license CC BY-NC-ND.

Contents

About the Editors vii

Veeriah Jegatheesan, Chettiyappan Visvanathan, Li Shu, Faisal I. Hai and Ludovic F. Dumée
Applications of Membranes for Sustainability
Reprinted from: *Membranes* **2021**, *11*, 629, doi:10.3390/membranes11080629 1

Arbab Tufail, William E. Price and Faisal I. Hai
Impact of Inorganic Ions and Organic Matter on the Removal of Trace Organic Contaminants by Combined Direct Contact Membrane Distillation–UV Photolysis
Reprinted from: *Membranes* **2020**, *10*, 428, doi:10.3390/membranes10120428 5

Susanthi Liyanaarachchi, Veeriah Jegatheesan, Li Shu, Ho Kyong Shon, Shobha Muthukumaran and Chun Qing Li
Evaluating the Feasibility of Forward Osmosis in Diluting RO Concentrate Using Pretreatment Backwash Water
Reprinted from: *Membranes* **2020**, *10*, 35, doi:10.3390/membranes10030035 21

Shruti Sakarkar, Shobha Muthukumaran and Veeriah Jegatheesan
Tailoring the Effects of Titanium Dioxide (TiO_2) and Polyvinyl Alcohol (PVA) in the Separation and Antifouling Performance of Thin-Film Composite Polyvinylidene Fluoride (PVDF) Membrane
Reprinted from: *Membranes* **2021**, *11*, 241, doi:10.3390/membranes11040241 39

Kajeephan Samree, Pen-umpai Srithai, Panaya Kotchaplai, Pumis Thuptimdang, Pisut Painmanakul, Mali Hunsom and Sermpong Sairiam
Enhancing the Antibacterial Properties of PVDF Membrane by Hydrophilic Surface Modification Using Titanium Dioxide and Silver Nanoparticles
Reprinted from: *Membranes* **2020**, *10*, 289, doi:10.3390/membranes10100289 65

Xianjun Du, Yaoke Shi, Veeriah Jegatheesan and Izaz Ul Haq
A Review on the Mechanism, Impacts and Control Methods of Membrane Fouling in MBR System
Reprinted from: *Membranes* **2020**, *10*, 24, doi:10.3390/membranes10020024 85

Yaoke Shi, Zhiwen Wang, Xianjun Du, Bin Gong, Veeriah Jegatheesan and Izaz Ul Haq
Recent Advances in the Prediction of Fouling in Membrane Bioreactors
Reprinted from: *Membranes* **2021**, *11*, 381, doi:10.3390/membranes11060381 119

About the Editors

Veeriah Jegatheesan is a Professor of Environmental Engineering at RMIT University, Melbourne, Australia. Jega is the founder and Chairman of the international conference series on Challenges in Environmental Science & Engineering (CESE), which has been held annually since 2008. Jega has conducted extensive research into various applications of membranes. He has over 450 publications, including more than 170 peer-reviewed journal articles and 5 edited books. In 2019, the Stormwater Industry Association (Australia) appointed him as one of the Governance Panel members for the Australian Stormwater Quality Improvement Device Evaluation Protocol (SQIDEP). Jega is the Editor-In-Chief of a book series entitled Applied Environmental Science and Engineering (AESE) for a Sustainable Future, published by Springer. Jega has been the Editor-in-Chief of the Environmental Quality Management journal (Wiley Publisher) since January 2020.

Chettiyappan Visvanathan is a Professor of the Environmental Engineering and Management Program, School of Environment, Resources and Development, Asian Institute of Technology. He has a Ph.D. in Chemical/Environmental Engineering from Institute National Polytechnique, Toulouse, France. His main areas of research interests include membrane technologies for water and wastewater treatment, cleaner production and solid waste disposal and management. In the field of membrane technology, his main research interests focus on the development of MBR systems (both aerobic and anaerobic), water recycling and industrial wastewater treatment. He has published more than 150 international journal papers and has more than 30 years of experience teaching environmental engineering and management-related courses at graduate level at AIT. His professional experiences include acting as a Project Engineer for the Asia Division, the International Training Center for Water Resources Management, Sophia Antipolis, France, and a being short-term consultant for the UNEP Industry and Environment Office, Paris, France.

Li Shu is an Adjunct Associate Professor at Edith Cowan University, Australia and the Managing Director of LJS Environment, Australia. She is a Guest Professor at Shandong Normal University. Li has more than 150 publications. She, along with her students and colleagues, was the first to document images of water clusters taken by a microscope. They also proposed the structure of water. With this proposed structure of water, water's mysterious behaviors could be explained. Li's research interest is in water and wastewater treatment using membranes, in addition to resource recovery and zero liquid discharge. She taught Water and Wastewater Systems to undergraduate and post graduate students. She is one of the founders of an International Conference "Challenges in Environmental Science and Technology", CESE, and has been the Co-Chair of the Conference since 2008. She is an Editorial Board Member and a Guest Editor of many prestigious journals.

Faisal I. Hai is the director of the Strategic Water Infrastructure Lab (SWIL) and the coordinator of the 'Environmental Engineering and Water Resources Powerhouse' of the University of Wollongong, Australia. He is an executive board member of the Membrane Society of Australasia. Professor Hai is an Editor of 'Water Science and Technology' (IWA, UK) and 'Journal of Water and Environment Technology' (Japan Society on Water Environment) and a member of the Editorial Board of numerous Elsevier journals. He is the lead editor of 'Membrane Biological Reactors', one of the bestsellers from IWA publishing, UK. A recognized authority in advanced wastewater treatment and reuse, Prof Hai has forged strong collaborations with industry and internationally leading researchers, which have led to collaborative grants and/or high-caliber publications.

Ludovic F. Dumée is an Assistant Professor of Separation Materials within the department of Chemical Engineering at Khalifa University, and is interested in the development and application of advanced separation materials, primarily focusing on membranes and catalysts. His research interests concern the understanding of nanoscale interactions between contaminants and surfaces, as well as the design of reactive and stimuli-responsive materials in the water, gas and healthcare applications. His team is interested in the engineering and combinatorial inclusion of carbon allotropes and ceramics nano-coatings across metal and polymeric materials to control the texture and reactivity of separation materials at nanoscale. He also focuses on the development of innovative solutions to tackle challenges arising from emerging micro-pollutants, such as PFAS or microplastics, as well as the simultaneous separation and conversion of gases into fuels and monomers.

Editorial

Applications of Membranes for Sustainability

Veeriah Jegatheesan [1,*], Chettiyappan Visvanathan [2], Li Shu [3,4], Faisal I. Hai [5] and Ludovic F. Dumée [6]

1. Water: Effective Technologies and Tools Research Centre and School of Engineering, RMIT University, Melbourne, VIC 3000, Australia
2. School of Environment, Resources and Development, Asian Institute of Technology, Klongluang, Pathumthani 12120, Thailand; visu@ait.ac.th
3. LJS Environment, Parkville, VIC 3052, Australia; li.shu846@gmail.com
4. School of Engineering, Edith Cowan University, Joondalup, WA 6027, Australia
5. Strategic Water Infrastructure Laboratory, School of Civil, Mining and Environmental Engineering, University of Wollongong, Wollongong, NSW 2522, Australia; faisal@uow.edu.au
6. Department of Chemical Engineering, Khalifa University, Abu Dhabi P.O. Box 127788, United Arab Emirates; ludovic.dumee@ku.ac.ae
* Correspondence: jega.jegatheesan@rmit.edu.au

Citation: Jegatheesan, V.; Visvanathan, C.; Shu, L.; Hai, F.I.; Dumée, L.F. Applications of Membranes for Sustainability. *Membranes* **2021**, *11*, 629. https://doi.org/10.3390/membranes11080629

Received: 6 August 2021
Accepted: 13 August 2021
Published: 16 August 2021

Publisher's Note: MDPI stays neutral with regard to jurisdictional claims in published maps and institutional affiliations.

Copyright: © 2021 by the authors. Licensee MDPI, Basel, Switzerland. This article is an open access article distributed under the terms and conditions of the Creative Commons Attribution (CC BY) license (https://creativecommons.org/licenses/by/4.0/).

Applications of membranes in water and wastewater treatment, desalination, as well as other purification processes, have become more widespread over the past few decades. However, we continue to ponder on (i) how well do we get maximum benefits from membrane applications? (ii) In what areas of membranes do we need to further our fundamental understanding? (iii) Which membrane technologies find difficulties in scale-up? (iv) How good are we in integrating membrane technologies? (v) How can we improve the reliability of membranes and reduce the cost? (vi) Where are we in disseminating the needs if we are to use membranes all over the world? (vii) How can we bring true global collaborations? The membrane community all over the world is continuing to contribute to answer those questions. This Special Issue is also contributing answers to some of the above questions. The call for papers for this Special Issue invited researchers broadly, and the delegates of the International Conference on the Challenges in Environmental Science and Engineering, CESE-2019 (3–7 November 2019, at the Grand Hi-Lai Hotel in Kaohsiung, Taiwan) in particular, who are working on the above to submit original research papers, critical review articles, case studies and technical notes.

Six articles investigating the performance of hybrid direct contact membrane distillation (DCMD)/UV photolysis and reverse osmosis (RO)/forward osmosis (FO) membranes, modification of membranes for photocatalysis and enhanced anti-bacterial properties, reviewing fouling of membrane bioreactors (MBRs) and predicting the fouling of MBRs are included in this collection. Tufail et al. [1] investigated the performance of hybrid DCMD and UV photolysis in degrading five trace organic contaminants (TrOCs). The study found that the nature and extent of the impact of inorganic ions, such as halides, nitrates and carbonates, as well as organic substance (humic acid), in degrading TrOCs depended on the type of TrOCs (phenolic contaminants such as bisphenol A and oxybenzone and non-phenolic contaminants such as sulfamethoxazole, carbamazepine, and diclofenac) and the concentration of the interfering ions. Thus, continued research is warranted to develop a database containing the performance of such a hybrid system in treating various TrOCs. Liyanaarachchi et al. [2] utilised FO to dilute RO concentrate, either to increase the water recovery from the RO system, or to discharge the diluted RO concentrate safely to the environment. They utilised the pre-treatment (sand filter) backwash water containing ferric hydroxide as the feed solution and RO concentrate as the draw solution for the FO process. The study found that cellulose triacetate (CTA) flat sheet FO membrane produced higher flux (3–6 L m^{-2} h^{-1}) compared to that produced by polyamide (PA) hollow fibre FO membrane (less than 2.5 L m^{-2} h^{-1}) under the same experimental conditions. Long-term studies conducted on the flat sheet FO membranes showed that fouling due to ferric hydroxide sludge did not allow the water flux to increase more than 3.15 L m^{-2} h^{-1}. Increasing the water flux needs further investigation in order for the process to have practical applications.

Sakakar et al. [3] synthesised thin-film composite (TFC) polyvinylidene fluoride (PVDF) membranes by coating with titanium dioxide (TiO$_2$)/polyvinyl alcohol (PVA) solution through the dip-coating method, and cross-linked it with glutaraldehyde to improve the thermal and chemical stability of the thin film coating. The study found that the layer of TiO$_2$ nanoparticles on the PVDF membranes reduced the fouling effects compared to the plain PVDF membrane. The study also showed that nearly 78% methyl orange and 47% reactive blue dyes were removed by the TFC membrane, along with photodegradation. However, the membrane can be damaged by the radicals formed by TiO$_2$ during UV irradiation. Further studies are required to improve the stability of the membrane by finding suitable polymers or using inorganic membranes which are less susceptible to the attack by the radicals. Samree et al. [4] modified the PVDF membrane by coating it with nanoparticles of titanium dioxide (TiO$_2$-NP) and silver (Ag-NP) at different concentrations and coating times to improve the hydrophilicity and antibacterial properties of the membrane. The study compared both the plain and modified PVDF membrane with respect to their performance in reducing Escherichia coli cells and inhibiting the formation of biofilm on the membrane surface. Compared to plain PVDF membrane, the modified membrane exhibited antibacterial efficiency up to 94% against E. coli cells, and inhibition up to 65% of the biofilm mass reduction.

Du et al. [5] carried out an extensive review on the mechanisms, impacts and control methods of membrane fouling in MBR systems. Compared to conventional activated sludge process, an MBR has many advantages, such as good effluent quality, small floor space, low residual sludge yield and easy automatic controls. It also allows slow growing microbes to exist and therefore remove hardly degradable pollutants. The treated effluent by MBR will generally meet Class A standards and therefore can be reused in various applications. However, membrane fouling is the main obstacle to the wider application of MBR. The authors suggest focussing the future research on the following: (i) Exploring real-time formation of organic fouling of membranes to optimise periodic cleaning. (ii) Developing new membranes with increased anti-fouling properties; low energy consumption, less chemical usage, and easier operation and maintenance should also be researched. Comprehensive economic analysis, life cycle assessment, and carbon footprint analysis of different direct membrane filtration processes should be conducted in order to identify the most suitable system configuration for further scale-up. (iii) Finding hybrid methods with low energy consumption for improving the application of ultrasonic technology in full-scale MBR systems. (iv) Optimising the aeration; the method of aeration could be optimized according to computational fluid dynamic modeling on the fluidization and the scouring behavior of the particles in MBRs. Moreover, the attachment tendency of biofilm colonizers on the medium and membranes should be assessed. Shi et al. [6] explored the recent advances in predicting the fouling of MBRs. They reviewed the techniques available to predict fouling in MBRs and discussed the problems associated with predicting fouling status using artificial neural networks and mathematical models. The authors suggest that the following need further research: (i) Fouling mechanisms in MBRs of different structures and scales with the development of accurate and real-time online data on membrane fouling. (ii) Study on remaining useful life (RUL) predictions of the membrane modules at various failure modes, as failure of membrane modules is usually caused by the synergistic effect of multiple failure modes. (3) Intelligent feature extraction by deep learning, such as a deep belief network and convolutional neural network.

Thus, the articles provided in this Special Issue disseminate both the current knowledge available and future research needs in the respective topics discussed. We hope this collection will be useful for developing plans for future research topics on the applications of membranes for sustainability.

Funding: This research received no external funding.

Acknowledgments: The guest editors are grateful to all the authors that contributed to this Special Issue.

Conflicts of Interest: The guest editors declare no conflict of interest.

References

1. Tufail, A.; Price, W.E.; Hai, F.I. Impact of Inorganic Ions and Organic Matter on the Removal of Trace Organic Contaminants by Combined Direct Contact Membrane Distillation–UV Photolysis. *Membranes* **2020**, *10*, 428. [CrossRef]
2. Liyanaarachchi, S.; Jegatheesan, V.; Shu, L.; Shon, H.K.; Muthukumaran, S.; Li, C.Q. Evaluating the Feasibility of Forward Osmosis in Diluting RO Concentrate Using Pretreatment Backwash Water. *Membranes* **2020**, *10*, 35. [CrossRef] [PubMed]
3. Sakarkar, S.; Muthukumaran, S.; Jegatheesan, V. Tailoring the Effects of Titanium Dioxide (TiO$_2$) and Polyvinyl Alcohol (PVA) in the Separation and Antifouling Performance of Thin-Film Composite Polyvinylidene Fluoride (PVDF) Membrane. *Membranes* **2021**, *11*, 241. [CrossRef] [PubMed]
4. Samree, K.; Srithai, P.-u.; Kotchaplai, P.; Thuptimdang, P.; Painmanakul, P.; Hunsom, M.; Sairiam, S. Enhancing the Antibacterial Properties of PVDF Membrane by Hydrophilic Surface Modification Using Titanium Dioxide and Silver Nanoparticles. *Membranes* **2020**, *10*, 289. [CrossRef] [PubMed]
5. Du, X.; Shi, Y.; Jegatheesan, V.; Haq, I.U. A Review on the Mechanism, Impacts and Control Methods of Membrane Fouling in MBR System. *Membranes* **2020**, *10*, 24. [CrossRef] [PubMed]
6. Shi, Y.; Wang, Z.; Du, X.; Gong, B.; Jegatheesan, V.; Haq, I.U. Recent Advances in the Prediction of Fouling in Membrane Bioreactors. *Membranes* **2021**, *11*, 381. [CrossRef]

Article

Impact of Inorganic Ions and Organic Matter on the Removal of Trace Organic Contaminants by Combined Direct Contact Membrane Distillation–UV Photolysis

Arbab Tufail [1], William E. Price [2] and Faisal I. Hai [1],*

[1] Strategic Water Infrastructure Laboratory, School of Civil, Mining and Environmental Engineering, University of Wollongong, Wollongong, NSW 2522, Australia; at742@uowmail.edu.au
[2] Strategic Water Infrastructure Laboratory, School of Chemistry and Molecular Bioscience, University of Wollongong, Wollongong, NSW 2522, Australia; wprice@uow.edu.au
* Correspondence: faisal@uow.edu.au; Tel.: +61-2-4221-3054

Received: 15 November 2020; Accepted: 14 December 2020; Published: 15 December 2020

Abstract: This study investigated the degradation of five trace organic contaminants (TrOCs) by integrated direct contact membrane distillation (DCMD) and UV photolysis. Specifically, the influence of inorganic ions including halide, nitrate, and carbonate on the performance of the DCMD–UV process was evaluated. TrOC degradation improved in the presence of different concentrations (1–100 mM) of fluoride ion and chloride ion (1 mM). With a few exceptions, a major negative impact of iodide ion was observed on the removal of the investigated TrOCs. Of particular interest, nitrate ion significantly improved TrOC degradation, while bicarbonate ion exerted variable influence—from promoting to inhibiting impact—on TrOC degradation. The performance of DCMD–UV photolysis was also studied for TrOC degradation in the presence of natural organic matter, humic acid. Results indicated that at a concentration of 1 mg/L, humic acid improved the degradation of the phenolic contaminants (bisphenol A and oxybenzone) while it inhibited the degradation of the non-phenolic contaminants (sulfamethoxazole, carbamazepine, and diclofenac). Overall, our study reports the varying impact of different inorganic and organic ions present in natural water on the degradation of TrOCs by integrated DCMD–UV photolysis: the nature and extent of the impact of the ions depend on the type of TrOCs and the concentration of the interfering ions.

Keywords: photodegradation; membrane distillation; halide ions; nitrate ions; carbonate ions; humic acid; trace organic contaminants

1. Introduction

Trace organic contaminants (TrOCs)—namely pharmaceuticals and personal care products, pesticides, surfactants, and industrial reagents—occur at the level of nanogram to microgram per litre in wastewater and polluted waterbodies [1–3]. The presence of these contaminants even in trace concentrations in the environment raises concern because of their harmful effects on human beings and aquatic lives [4]. A systematic analysis of the available studies shows that conventional wastewater treatments are not capable of effective elimination of persistent TrOCs. This results in their occurrence in the surface water and groundwater [5–7]. Therefore, an effective treatment process is essential for TrOC removal.

Membrane distillation involves moderate-temperature distillation compared to conventional distillation processes (e.g., steam distillation). In this process, water is transported in vapour form from the feed side to distillate through porous hydrophobic membranes. Among the available configurations, direct contact membrane distillation (DCMD) is a promising technology [8–10]. This is a high-retention membrane process that involves low fouling potential compared to pressure-driven membrane processes and can be less energy-intensive when low-grade waste heat is used [11,12]. A few recent studies have reported the removal of TrOCs by this process [13–15]. For TrOC removal, a temperature difference of 15–20 °C is created to drive the transport of water vapour from feed to distillate. Non-volatile TrOCs can be effectively removed by the DCMD system as mass transport occurs in the vapour phase [16,17]. The TrOCs effectively retained by DCMD (i.e., the membrane concentrate) require additional treatment before disposal to the environment [18]. We propose that the DCMD system can be integrated with UV photolysis, thereby devising the DCMD–UV photolysis process, where the DCMD membrane can retain the contaminants and UV photolysis can continuously degrade them.

More commonly applied in water disinfection, UV photolysis can also be applied for the degradation of TrOCs. The literature shows that contaminants with photolabile moieties (e.g., -Cl, -OH) absorb photon energy from UV light and undergo oxidation [19–21]. There is limited information on overall TrOC removal by combined DCMD–UV photolysis. Mozia et al. [22] investigated diclofenac removal by combined DCMD and UV photolysis and revealed its complete elimination within 4 h. Another study reported the removal of TrOCs by DCMD preceding UV photolysis and showed 27–88% removal depending on the TrOC [18].

DCMD can effectively retain the non-volatile TrOCs and UV photolysis can degrade them with TrOC-specific efficiency. However, the presence of inorganic ions—namely halide ions, nitrate ions, and bicarbonate ions—in water can influence TrOC removal efficiency [20,23,24]. For instance, Li et al. [25] demonstrated the effect of bromide and chloride ions on the photodegradation of three antibiotics (sulfamethoxazole, sulfamethazine, and sulphapyridine) and reported improved rate constants for sulfamethazine and sulphapyridine in the presence of chloride ion while the rate constants decreased for sulfamethoxazole. In that study, the rate constant for all the TrOCs decreased in the presence of bromide ion. Yang et al. [26] reported the inhibitory effect of bicarbonate ion on sulfamethoxazole photodegradation. Chloride and nitrate ions also inhibited the photodegradation of diclofenac, while bicarbonate ion enhanced diclofenac degradation [26]. The efficiency of the DCMD–UV treatment may be also affected in the presence of natural organic matter such as humic acid in water. The presence of humic acid gives water a yellowish-brown colour which can reduce UV light penetration. Humic acid also acts as an OH radical scavenger. However, the literature illustrates contrasting results regarding the effect of humic acid on the UV photolysis of TrOCs. For instance, the degradation rate constant of 17α-ethinylestradiol and 17β-estradiol greatly improved in the presence of humic acid [27,28] while naproxen degradation was inhibited [29]. Overall, information on the impact of co-occurring ions in water on TrOC degradation by DCMD–UV is very limited.

This study investigated TrOC retention by DCMD and their degradation by combined DCMD–UV photolysis using five selected TrOCs including bisphenol A, oxybenzone, diclofenac, carbamazepine, and sulfamethoxazole, which have diverse physicochemical properties and are commonly detected in wastewater and polluted waterbodies. Because the presence of organic and inorganic ions in water may significantly influence TrOC degradation, the effect of inorganic ions such as halide, nitrate, and bicarbonate ions as well as natural organic matter (i.e., humic acid) on DCMD–UV photolysis was systematically studied. This is the first study to elucidate the effect of different interfering ions on TrOC removal from water by the integrated DCMD–UV system.

2. Materials and Methods

2.1. Materials

In this study, five TrOCs including bisphenol A, oxybenzone, diclofenac, carbamazepine, and sulfamethoxazole were selected in view of their common detection in wastewater-affected natural waterbodies [1]. Table 1 illustrates the main physicochemical characteristics of these TrOCs, namely contaminant structures, hydrophobicity, and dissociation constants. All tested TrOCs were bought from Sigma Aldrich (Castle Hill, Australia) and had greater than 98% purity. Inorganic salts (i.e., sodium fluoride, sodium chloride, sodium bromide, sodium iodide, sodium nitrate, and sodium bicarbonate), humic acid, methanol, and HPLC-grade acetonitrile were also sourced from the same supplier. Ultrapure Milli-Q water (Millipore S.A.S, Molsheim, France) was used in all experiments and each experiment was conducted in duplicate. According to the supplier, Milli-Q water has a resistivity of 18.2 MΩ·cm and a total organic carbon (TOC) of less than 5 ppb.

Table 1. Physicochemical properties of the selected trace organic contaminants.

Compound	Molecular Weight (g/mol)	Log D at pH 7	Vapour Pressure (mmHg)	pK_H at pH 7
Bisphenol A	228.29	3.64	5.34×10^{-7}	8.66
Sulfamethoxazole	253.28	−0.22	1.52×10^{-12}	11.81
Diclofenac	296.15	1.77	1.59×10^{-7}	8.68
Oxybenzone	228.24	3.89	5.26×10^{-6}	7.80
Carbamazepine	236.27	1.89	5.78×10^{-7}	9.08

Note: Chemical structure, molecular weight, Log D, vapour pressure, and pKH values were taken from SciFinder Scholar.

2.2. Sample Preparation

A stock solution of the selected TrOCs was prepared by adding the TrOCs each at a concentration of 2 g/L to pure methanol. The stock solution was stored in the dark at −18 °C and used within one month. A working solution was freshly prepared by diluting the stock solution in Milli-Q water to obtain 1 mg/L concentration of the TrOCs. A calibration curve of each TrOC was established in the range 0.1–1 mg/L to quantify TrOC concentration in samples by HPLC. Stock solutions (1 M) were prepared for each inorganic salt. Humic acid stock solution was at a concentration 1 g/L. These solutions were further diluted to obtain working solutions of the salts (1 5, 10, and 100 mM) and humic acid (1, 5, and 10 mg/L).

2.3. DCMD and UV Setup and Operation Protocol

A lab-scale membrane distillation rig comprising a direct contact membrane cell and a reactor made of glass, as shown in Figure 1, was used to conduct experiments. The liquid in the glass reactor was used as feed for the DCMD module. The feed tank (working volume of 5 L) was placed in a temperature controlled (30 ± 1 °C) water bath. The water bath was equipped with a heating unit (Julabo, Seelbach, Germany) to keep the feed temperature at 30 ± 1 °C. Temperature for distillate was set up at 10 ± 1 °C using a chiller (SC100-A10, Thermo Scientific, Vernon Hills, IL, USA). The setup was designed and operated following a previously published protocol [16]. The initial feed volume was 1.5 L and the nominal concentration of each TrOC in the feed solution was 1 mg/L. The impact of different concentrations of each ion (1, 5, 10, and 100 mM) and humic acid (1, 5, and 10 mg/L) was studied in separate runs. In each run, the DCMD–UV setup was run for 60 min. The DCMD system was run in recirculation mode and flow rate was kept at 1 L/min for both feed and permeate. The permeate flux was recorded every 5 min and the DCMD system was initially operated for 60 min to verify complete retention of the selected TrOCs by DCMD.

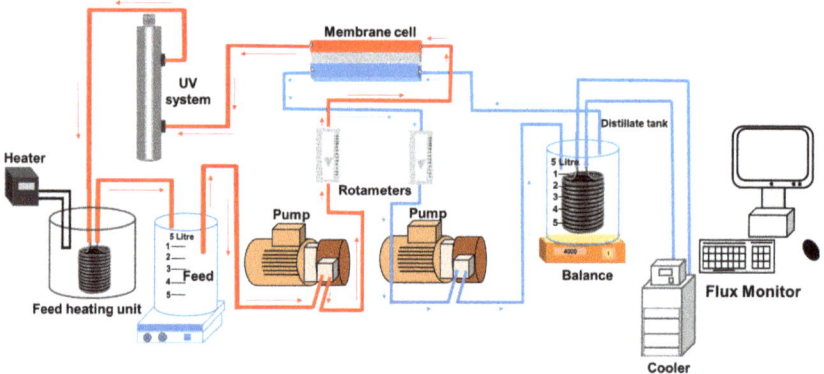

Figure 1. Experimental setup for integrated direct contact membrane distillation–UV photolysis treatment.

The DCMD module was made up of acrylic glass. Feed and permeate flow channels (145 × 95 × 3 mm) were engraved on each block. Polytetrafluoroethylene (PTFE) membrane was purchased from Ningbo Porous Membrane Technology (Ningbo, China) and was used in this study. The PTFE membrane was hydrophobic in nature, with a surface area, nominal pore size, thickness, and porosity of 221 cm^2, 0.2 µm, 60 µm, and 80%, respectively.

A bench-scale UV oxidation setup (UVG SLT30 model) purchased from UV Guard (Castle Hill, NSW, Australia) and was integrated with the DCMD setup as shown in Figure 1. It had a working volume of 1.1 L. It comprised an outer 316-grade stainless steel housing protecting an inner quartz reactor. The principal wavelength of the 60 cm long UV lamp (30 W) was 254 nm. According to the supplier, when operated at a flowrate of 16 L/min, this setup provides a UV dose of 40 mJ/cm^2 (as calculated using UVCalc® software based on a UV transmittance of 85%). With a flowrate of 1 L/min in this study, the estimated UV dose was around 750 mJ/cm^2. The lamp was placed inside the quartz reactor. It provided continuous exposure to the test solution present inside the outer reactor. All experiments combining DCMD–UV treatment were conducted for a UV exposure time of 60 min.

In this work, the removal of each TrOC in the DCMD–UV system was calculated by establishing a mass balance of TrOC concentration in feed and permeate at the start and end of each run [30]. This is the first study combining DCMD and UV treatment for the removal of TrOCs from their mixture and assessing the effect of inorganic ions (e.g., halide, nitrate, and carbonate) and humic acid on TrOC removal by this process.

2.4. TrOC Analysis

The concentrations of TrOCs present in the samples were measured using an HPLC system (Shimadzu, Kyoto, Japan) following a previously published protocol [24]. The limit of quantification for the TrOCs was 10 µg/L. The accuracy of quantification was always confirmed by running standard solutions. Removal of TrOCs was calculated as R (%) = $(1 - \frac{C_t}{C_0}) \times 100$, where C_0 and C_t are initial mass and mass at time of sampling, respectively.

3. Results

3.1. Results and Discussion

TrOC Removal by DCMD

In DCMD, water passes through the hydrophobic membrane in vapour form. Retention of TrOCs in the feed reactor depends on their volatility and distribution coefficient (log D). TrOCs with pK_H value greater than 9 have low volatility and are expected to be well removed by the DCMD system. Log D represents hydrophobicity and can also affect the transport of TrOCs through the MD membrane [17].

Interestingly, Figure 2 shows that irrespective of Log D and pK_H values, greater than 99% retention of the TrOCs was achieved by the DCMD system. Previously, Wijekoon et al. [17] reported 81% removal for oxybenzone while > 97% rejection for bisphenol A, carbamazepine, and diclofenac at feed and distillate temperatures of 40 and 10 °C, respectively. This difference in TrOC removal can be attributed to the lower feed-side temperature (i.e., 30 vs. 40 °C) used in the current study, which would have reduced the chance of TrOCs escaping in vapor form.

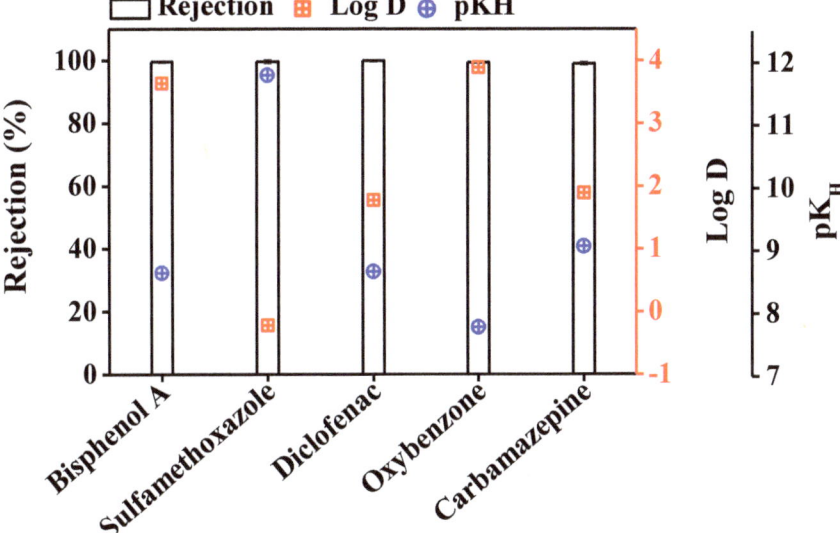

Figure 2. Retention of trace organic contaminants in DCMD system. Retention % was evaluated by establishing a mass balance of trace organic contaminants in the feed at the start and end of the experiment. The DCMD system retained the trace organic contaminants completely. Operating conditions for DCMD system: temperature of the feed and the distillate was set at 30 and 10 °C, respectively; cross flow rate for feed and distillate was maintained at 1 L/min. Error bars represent the standard deviation of duplicate samples.

In our previous study [18], where the DCMD system was run for 18 h (compared to 1 h in the current study), similar TrOC retention was observed with negligible flux decline. Fouling and wetting of membrane can significantly affect TrOC retention when the feed water contains other impurities than TrOC. Future studies are suggested to shed light on this aspect. However, this is beyond the scope of the current study.

3.2. Fate of TrOCs in DCMD–UV Photolysis Process

TrOCs retained by the DCMD process eventually accumulate in the feed reactor. This requires additional treatment of the DCMD concentrate before disposal into the environment. Thus, DCMD was combined with UV photolysis to simultaneously retain and degrade TrOCs.

The DCMD permeate, i.e., the treated final effluent, was already virtually TrOC-free. The removal efficiency by DCMD–UV discussed in this section refers to the reduction of the concentration of retained TrOCs in feed solution by UV degradation. Samples taken from the feed side revealed substantial degradation of sulfamethoxazole (87%), bisphenol A (95%), and diclofenac (71%) but rather limited degradation of carbamazepine (9%) and oxybenzone (22%) (Figure 3). High removal of sulfamethoxazole, bisphenol A, and diclofenac can be attributed to the presence of more than one photolabile functional group (-CH3, -OH, -COOH, -NH) in their structures, which makes them less stable in the presence of UV irradiation [19]. Low removal of oxybenzone despite the presence of an –OH group in its molecule can be attributed to the presence of fewer photolabile functional groups and a stable benzene ring. Carbamazepine is resistant to photodegradation due to the absence of photosensitive functional groups [31].

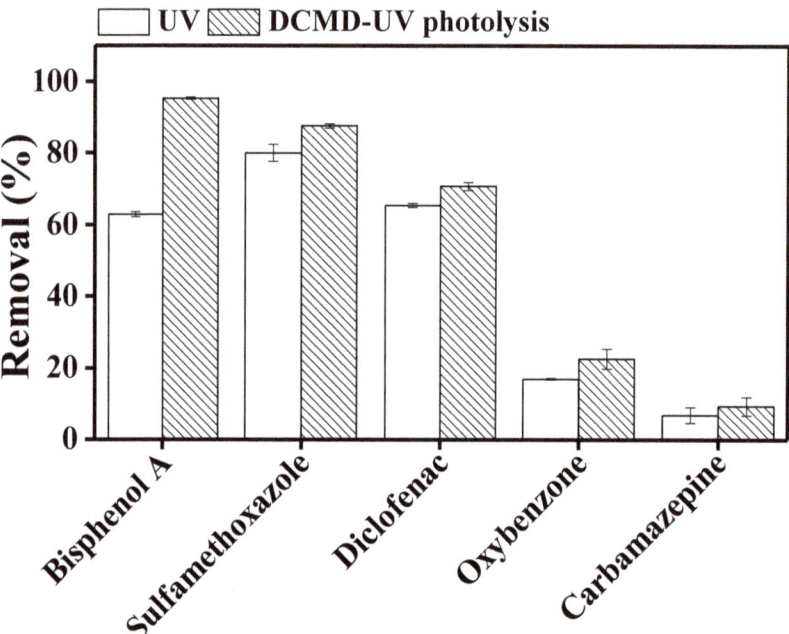

Figure 3. Degradation of trace organic contaminants in UV photolysis and integrated DCMD–UV photolysis. Operating conditions for DCMD: temperature of feed and distillate was set at 30 and 10 °C, respectively; cross flow rate for feed and distillate was maintained at 1 L/min. Operating conditions for photolysis: UV dose was around 750 mJ/cm^2 and the reaction time was 60 min. Error bars represent the standard deviation of duplicate samples.

Our observation regarding carbamazepine and oxybenzone is consistent with previous studies which reported low removal of these TrOCs by direct UV photolysis at 254 nm [31,32]. Mozia et al. [22] reported 79–96% diclofenac degradation at different initial concentrations of the TrOC, which is consistent with the degradation range that we have observed for this compound. On the other hand, while exploring UV post treatment of DCMD concentrate, Tufail et al. [18] reported lower degradation of bisphenol A compared to the current study. In the current study, the UV system was integrated with the DCMD system. On the other hand, Tufail et al. [18] first operated DCMD independently and then treated the membrane concentrate by UV. Thus, in the current study, the TrOC concentration in the test solution which was exposed to UV was one third. This may be the reason for the better bisphenol A removal.

We compared TrOC degradation in feed solution by UV when the UV system was operated separately versus when DCMD and UV were integrated in the same loop. While for the other TrOCs, UV degradation performance was similar irrespective of the arrangement of the UV and DCMD components, bisphenol A degradation by UV was significantly higher when DCMD and UV were integrated in the same loop (Figure 3). Two previous studies by Mozia et al. [22,33] reported the benefit of integrating the DCMD and UV photolysis in general: the DCMD system retains TrOCs in the feed reactor and increases their concentration, and UV photolysis results in the degradation of the TrOCs; however, they did not discuss the impact of the arrangement of the UV and DCMD components. Nevertheless, noting that in photocatalytic membrane reactors, combining membrane and UV in the same tank results various synergistic advantages [34], it would be interesting to further investigate this aspect. However, this is beyond the scope of the current study.

3.3. Effect of Inorganic Ions on the TrOC Removal by DCMD–UV Photolysis

3.3.1. Effect of Nitrate Ion

Upon UV irradiation, nitrate ion produces nitrite and hydroxyl radicals in the reaction system (Equations (1)–(3)) that may help in degrading contaminants.

$$NO_3^- \rightarrow [NO_3^-]^* \tag{1}$$

$$[NO_3^-]^* \rightarrow NO_2^- + O \tag{2}$$

$$[NO_3^-]^* \, NO_2^\bullet + O^{\bullet-} \rightarrow NO_2 + {}^\bullet OH + OH^- \tag{3}$$

Figure 4 shows a significant increase in TrOC degradation when nitrate ion is added. Furthermore, within the nitrate concentration range of 1–10 mM, TrOC degradation either remained unchanged (bisphenol A, sulfamethoxazole, and diclofenac) or increased gradually (oxybenzone and carbamazepine). The degradation of bisphenol A, oxybenzone, and carbamazepine reduced significantly at a nitrate concentration of 100 mM.

Consistent with our observation regarding sulfamethoxazole, Hao et al. [35] reported a promoting effect of nitrate ion on UV degradation of sulphonamide compounds. Similarly, as shown in Figure 4, removal of diclofenac increased consistently with increasing nitrate ion concentration (0–100 mM). This result is in line with the performance reported by Koumaki et al. [36]. On the other hand, in the current study, bisphenol A degradation reduced by 50% when the nitrate concentration was increased from 10 to 100 mM. Reduced degradation was also observed for carbamazepine and oxybenzone at this level of nitrate concentration (Figure 4). This can be attributed to the "shielding effect" due to the presence of high nitrate ion concentration in the reaction mixture [36]. Nitrate ion also absorbs UV light and, when present in excessive concentrations, this ion can create competition for the available number of photons [37]. The addition of nitrate ion can produce hydroxyl radicals, but these radicals play a small part in the whole degradation process when direct photodegradation plays the key role in TrOC degradation. Thus, TrOC degradation—for example, that of sulfamethoxazole—did not increase proportionately with nitrate ion concentration.

Overall, our results highlight the concentration-specific impact of nitrate ion on UV degradation of TrOCs.

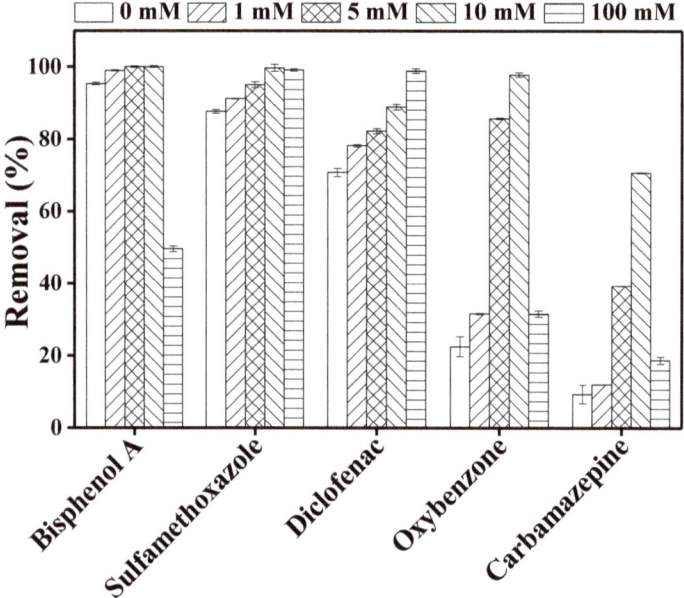

Figure 4. Impact of different concentration (0, 1, 5, 10, 100 mM) of nitrate ions on the degradation of trace organic contaminants in the integrated DCMD–UV photolysis. Permeate flux was 3.6 L/m^2·h, conductivity was around 4 µS/cm, and TrOC concentration in permeate was below the detection limit. Other operating conditions for integrated DCMD–UV photolysis are given in the caption of Figure 3.

3.3.2. Effect of Halide Ions

Halide ions—namely fluoride, chloride, bromide, and iodide ions—are ubiquitously detected in seawater, surface water, or groundwater over a wide concentration range of 0.1–500 mg/L [25,38]. These ions can undergo photoexcitation by UV radiation having a wavelength below 260 nm. In the presence of UV radiation, halide ions can produce radicals that selectively attack contaminants and degrade them [38]. Redox potential for fluoride, chloride, bromide, and iodide are 2.9, 2.59, 2.04, and 1.37 V, respectively [26,39]. On the other hand, halide ions act as hydroxyl radical scavengers by reacting with hydroxyl radicals to produce halide radicals, which reduces hydroxyl radical-mediated degradation of TrOCs [38,39].

In this study, with a few exceptions, bisphenol A removal significantly decreased in the presence of chloride, bromide, and iodide ions (Figure 5). The reduction in BPA degradation in the presence of halide ions can be attributed to the competition between BPA and the halide ions for UV irradiation. Moreover, the presence of halide ions may generate some other radicals, e.g., HOX, which may inhibit TrOC degradation [40]. For instance, Grebel et al. [41] reported a 90% reduction in the photodegradation of 17β-estradiol in the presence of 0.54 M chloride ion. They attributed this reduction to the ionic strength effect of chloride ions.

Sulfamethoxazole and diclofenac removal was mostly unaffected by the halide ions except for iodide ion. This is in line with previous studies where sulfamethoxazole degradation remained unchanged in the presence of chloride and bromide ions. For example, Li et al. [25] reported that chloride ion (0.54 M) and bromide ion (0.8 mM) did not affect the degradation of sulfapyridine and sulfamethoxazole.

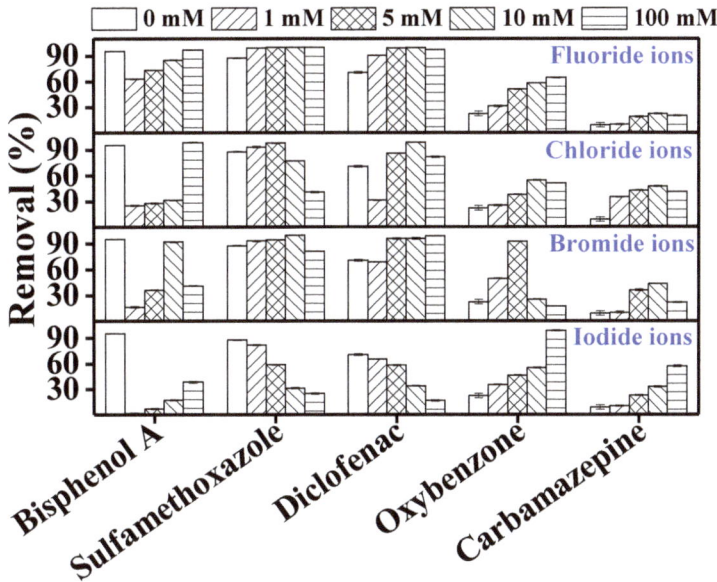

Figure 5. Impact of halide ions (fluoride, chloride, bromide, and iodide) on the degradation of trace organic contaminants in the integrated DCMD–UV photolysis. Permeate flux was 3.7 L/m^2·h, conductivity was around 4 µS/cm, and TrOC concentration in permeate was below the detection limit. Operating conditions for integrated DCMD–UV photolysis are given in the caption of Figure 3. Halide ion concentrations were 0, 1, 5, 10, 100 mM.

Sulfamethoxazole and diclofenac may involve a triplet-excited state in their photodegradation [25]. Halide ions could quench the triplet-excited state of TrOCs by the formation of complex intermediates between halide ions and the excited state of the TrOC, thus reducing their photodegradation. However, Li et al. [25] suggested that the oxidation potential of sulfamethoxazole at triplet-excited state is not large enough to react with halide ions and form intermediates. Therefore, except iodide, the presence of halide ions did not affect the degradation of sulfamethoxazole. Several studies have demonstrated that the presence of organic matter may create competition for UV light, thus affecting the photodegradation of TrOCs. A stronger light attenuation effect occurs in the presence of organic species with higher absorbance [39,41,42]. Among halide ions, iodide ion shows the highest absorbance that increases with its concentration. Therefore, inhibition of the TrOC degradation can be attributed to the attenuation effect of the halide ions [42].

Except for the significantly reduced removal in the presence of bromide ion beyond a concentration of 5 mM, oxybenzone removal gradually increased with halide concentrations. Oxybenzone can react with chloride ions and generate chlorinated by-products such as chloroform and halogenated methoxyphenols [43]. Therefore, conversion of oxybenzone increased in the presence of chloride ions. Other halide ions also form radicals in the presence of UV irradiation that may react with oxybenzone and promote or inhibit its conversion/degradation [44].

Compared to the other TrOCs, carbamazepine removal in the control experiments (i.e., in the absence of any halides) was originally much lower. Its removal slightly increased in the presence of the halides.

Notably, Li et al. [42] investigated the photodegradation of ibuprofen in the presence of different halide ions. The authors reported that, among all the halides, iodide ion shows the maximum light attenuation effect and can significantly impact the degradation of contaminants. In our study too, overall, with a few exceptions, a major negative impact of iodide ion was observed on the removal of the investigated TrOCs, which can be attributed to its maximum light attenuation effect as discussed above.

3.3.3. Effect of Bicarbonate Ion

Inorganic carbon occurs in water in the form of either carbonate or bicarbonate ions and can affect TrOC photodegradation efficiency [45]. TrOC degradation may reduce because this ion may shield UV radiation [36]. Bicarbonate ion may also scavenge hydroxyl radicals (Equation (4)) and generate carbonate radicals (strong one-electron oxidants): because of this, while the dissipated hydroxyl radical cannot take part in TrOC degradation, the carbonate radicals generated can selectively oxidize TrOCs [26,46].

$$OH + HCO_3^- \rightarrow H_2O + CO_3^- \quad (4)$$

Carbonate radicals react via electron transfer or hydrogen transfer with TrOCs having aromatic amines, thiols, and phenol groups. Only a handful of studies have reported the effect of bicarbonate on the photodegradation of contaminants [26]. In this study, except for oxybenzone, which showed an opposite trend, TrOC removal decreased significantly with the increase in bicarbonate concentration (Figure 6). Our observation is consistent with that of Mozia et al. [33], who reported reduced ibuprofen degradation by MD photocatalysis due to bicarbonate ion. Oxybenzone contains a phenolic functional group and thus its improved photodegradation in our study in the presence of bicarbonate ion can be attributed to the formation of carbonate radicals that degrade phenolic moieties via electron transfer. On the other hand, as in this study, Yang et al. [26] reported a reduction in sulfamethoxazole removal in the presence of bicarbonate ion at a concentration of 50 mM. A similar decreasing trend was observed for the photodegradation of bisphenol A and diclofenac in previous studies [47,48]. It is worth mentioning that in this study, TrOC removal inhibition was not significant at the lowest bicarbonate concentration tested (1 mM). Thus, it is possible that light attenuation at higher bicarbonate concentrations was one of the reasons for the deteriorated photodegradation of the TrOCs [42].

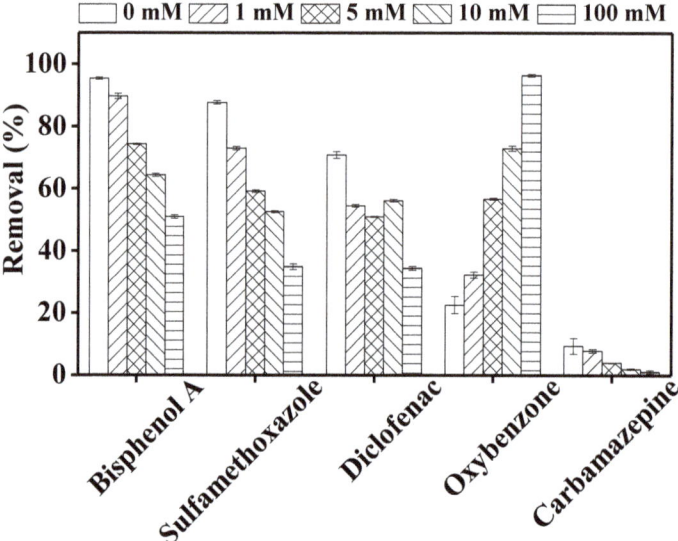

Figure 6. Impact of different concentrations (0, 1, 5, 10, 100 mM) of bicarbonate ions on the degradation of trace organic contaminants in the integrated DCMD–UV photolysis. Permeate flux was 3.5 L/m²·h, conductivity was around 4 µS/cm, and TrOC concentration in permeate was below the detection limit. Other operating conditions for integrated DCMD–UV photolysis are given in the caption of Figure 3.

3.4. Effect of Humic Acid on TrOC Removal by DCMD–UV Photolysis

Among the natural organic matter in water is humic acid, which has an average molecular weight of 2000–5000 and consists of a large portion of oxygen-containing functional groups. Humic acid is chromophoric in nature; thus, it is excited by UV irradiation having a wavelength ranging between 300 and 500 nm [29,49]. In general, humic acid can promote contaminant degradation through the generation of reactive oxygen species or retard degradation by shielding the UV radiation [39,49]. In the presence of sunlight (which emits predominantly UVA but also UVC), humic acid undergoes excitation and generates various radicals, namely hydroxyl radical ($^\bullet$OH), peroxy radical (ROO$^\bullet$), and singlet oxygen species ($O_2^{\bullet-}$) [36]. These radicals can attack TrOCs in the solution and oxidise them. By contrast, humic acid can also attenuate light as it can shield UV radiation.

Figure 7 illustrates the photodegradation of the investigated TrOCs in the presence of different concentrations (1, 5, 10 mg/L) of humic acid. With the exception of diclofenac, the impact of humic acid on TrOC removal was low. Calza et al. [49] mentioned that the excited triple state of humic acid plays an important role in the degradation of the phenolic contaminants. Both oxybenzone and bisphenol A contain hydroxyl functional groups and thus their degradation was not reduced in the presence of humic acid. On the other hand, in good agreement with our study, Zhang et al. [50] observed a slight inhibitory effect of humic acid (20 mg/L) on the photodegradation of sulfamethoxazole. Similarly, Wang et al. [51] observed reduced degradation of carbamazepine in the presence of humic acid. Consistent with our observation, previously, Koumaki et al. [36] reported that the degradation of diclofenac was greatly reduced in the presence of 20 mg/L of humic acid during 15 h of solar irradiation. Evidently, diclofenac removal is relatively more affected by the light attenuation in the presence of humic acid. In this study, sulfamethoxazole and diclofenac were well removed by direct photolysis but the degradation of diclofenac was greatly affected by humic acid. This difference in degradation can be attributed to their different molar absorption coefficients and pKa values [52,53].

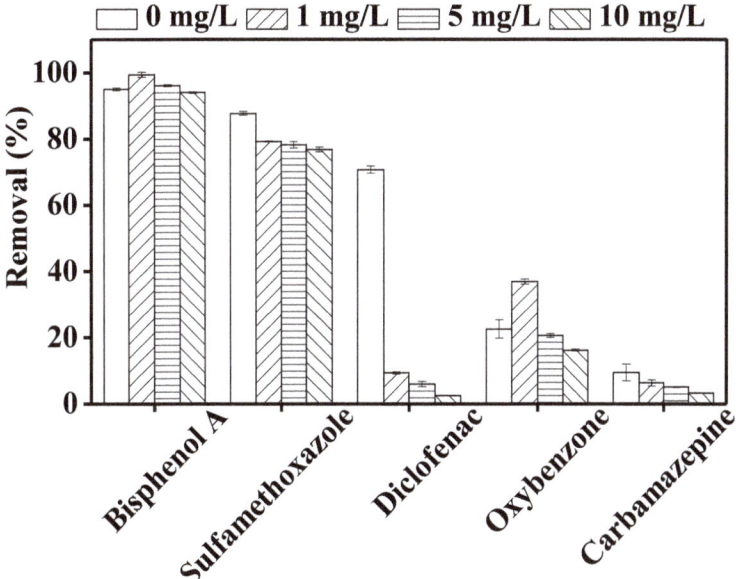

Figure 7. Impact of 1, 5, and 10 mg/L of humic acid on the degradation of trace organic contaminants in the integrated DCMD–UV photolysis. Operating conditions for integrated DCMD–UV photolysis are given in the caption of Figure 3.

4. Conclusions

In this study, we compared the degradation of five trace organic contaminants by UV photolysis and combined DCMD–UV photolysis. Results showed that all five investigated TrOCs were effectively retained (>99%) by the DCMD process, and the retained TrOCs were degraded by UV photolysis with TrOC-specific efficiency. TrOC degradation capacity by the integrated DCMD–UV photolysis process in the presence of humic acid and inorganic ions—namely halide, nitrate, and bicarbonate—was investigated. The nature and extent of the impact of the ions were observed to depend on the type of TrOCs and the concentration of the interfering ions. At a concentration of 1 mg/L, humic acid improved the degradation of the phenolic contaminants (bisphenol A and oxybenzone) while it inhibited the degradation of the non-phenolic contaminants (sulfamethoxazole, carbamazepine, and diclofenac). Conversely, the presence of a high concentration (10 mg/L) of humic acid overall inhibited the degradation of the TrOCs. With an exception, a major negative impact of iodide ion was observed on the removal of the investigated TrOCs. Of particular interest, fluoride and nitrate ions significantly improved TrOC degradation, while bicarbonate ion illustrated a variable influence—from promoting to inhibiting impact—on TrOC degradation.

Author Contributions: Conceptualization, methodology, formal analysis, investigation, F.I.H., W.E.P. and A.T.; resources, F.I.H. and W.E.P.; data curation, writing—original draft preparation, F.I.H., W.E.P. and A.T.; writing—review and editing, visualization, F.I.H., W.E.P. and A.T.; supervision, project administration, F.I.H. and W.E.P.; funding acquisition, F.I.H. All authors have read and agreed to the published version of the manuscript.

Funding: This research was funded by the UOW- HEC joint scholarship to Arbab Tufail.

Acknowledgments: The authors thank Faculty of Engineering and Information Sciences, UOW for access to the experimental setups and the analytical equipment.

Conflicts of Interest: The authors declare no conflict of interest.

References

1. Luo, Y.; Guo, W.; Ngo, H.H.; Nghiem, L.D.; Hai, F.I.; Zhang, J.; Liang, S.; Wang, X.C. A review on the occurrence of micropollutants in the aquatic environment and their fate and removal during wastewater treatment. *Sci. Total Environ.* **2014**, *473–474*, 619–641. [CrossRef]
2. Jelić, A.; Gros, M.; Petrović, M.; Ginebreda, A.; Barceló, J. Occurrence and Elimination of Pharmaceuticals during Conventional Wastewater Treatment. *Emerg. Prior. Pollut. Rivers* **2012**, *24*, 1–23. [CrossRef]
3. Hai, F.I.; Nghiem, L.D.; Khan, S.J.; Asif, M.B.; Price, W.E.; Yamamoto, K. Removal of Emerging Trace Organic Contaminants (TrOC) by MBR. In *Membrane Biological Reactors: Theory, Modeling, Design, Management and Applications to Wastewater Reuse*; IWA Publishing: London, UK, 2019; pp. 413–468.
4. Bayen, S. Occurrence, bioavailability and toxic effects of trace metals and organic contaminants in mangrove ecosystems: A review. *Environ. Int.* **2012**, *48*, 84–101. [CrossRef]
5. Tadkaew, N.; Hai, F.I.; McDonald, J.A.; Khan, S.J.; Nghiem, L.D. Removal of trace organics by MBR treatment: The role of molecular properties. *Water Res.* **2011**, *45*, 2439–2451. [CrossRef]
6. Wijekoon, K.C.; Hai, F.I.; Kang, J.; Price, W.E.; Guo, W.; Ngo, H.H.; Nghiem, L.D. The fate of pharmaceuticals, steroid hormones, phytoestrogens, UV-filters and pesticides during MBR treatment. *Bioresour. Technol.* **2013**, *144*, 247–254. [CrossRef] [PubMed]
7. Ikhlaq, A.; Kazmi, M.; Tufail, A.; Fatima, H.; Joya, K.S. Application of peanut shell ash as a low-cost support for Fenton-like catalytic removal of methylene blue in wastewater. *Desalin. Water Treat.* **2018**, *111*, 338–344. [CrossRef]
8. Chew, N.G.P.; Zhang, Y.; Goh, K.; Ho, J.S.; Xu, R.; Wang, R. Hierarchically Structured Janus Membrane Surfaces for Enhanced Membrane Distillation Performance. *ACS Appl. Mater. Interfaces* **2019**, *11*, 25524–25534. [CrossRef] [PubMed]
9. Chen, X.; Vanangamudi, A.; Wang, J.; Jegatheesan, J.; Mishra, V.; Sharma, R.; Gray, S.R.; Kujawa, J.; Kujawski, W.; Wicaksana, F.; et al. Direct contact membrane distillation for effective concentration of perfluoroalkyl substances—Impact of surface fouling and material stability. *Water Res.* **2020**, *182*, 116010. [CrossRef] [PubMed]

10. Couto, C.F.; Amaral, M.C.S.; Lange, L.C.; Santos, L.V.D.S. Effect of humic acid concentration on pharmaceutically active compounds (PhACs) rejection by direct contact membrane distillation (DCMD). *Sep. Purif. Technol.* **2019**, *212*, 920–928. [CrossRef]
11. Alkhudhiri, A.; Darwish, N.A.; Hilal, N. Membrane distillation: A comprehensive review. *Desalination* **2012**, *287*, 2–18. [CrossRef]
12. Wijekoon, K.C.; Hai, F.I.; Kang, J.; Price, W.E.; Guo, W.; Ngo, H.H.; Cath, T.Y.; Nghiem, L.D. A novel membrane distillation–thermophilic bioreactor system: Biological stability and trace organic compound removal. *Bioresour. Technol.* **2014**, *159*, 334–341. [CrossRef] [PubMed]
13. Chew, N.G.P.; Zhao, S.; Wang, R. Recent advances in membrane development for treating surfactant- and oil-containing feed streams via membrane distillation. *Adv. Colloid Interface Sci.* **2019**, *273*, 102022. [CrossRef] [PubMed]
14. Yeszhanov, A.; Korolkov, I.V.; Gorin, Y.G.; Dosmagambetova, S.S.; Zdorovets, M.V. Membrane distillation of pesticide solutions using hydrophobic track-etched membranes. *Chem. Pap.* **2020**, *74*, 3445–3453. [CrossRef]
15. Chew, N.G.P.; Zhao, S.; Loh, C.H.; Permogorov, N.; Wang, R. Surfactant effects on water recovery from produced water via direct-contact membrane distillation. *J. Membr. Sci.* **2017**, *528*, 126–134. [CrossRef]
16. Asif, M.B.; Hai, F.I.; Kang, J.; Van De Merwe, J.P.; Leusch, F.D.; Price, W.E.; Nghiem, L.D. Biocatalytic degradation of pharmaceuticals, personal care products, industrial chemicals, steroid hormones and pesticides in a membrane distillation-enzymatic bioreactor. *Bioresour. Technol.* **2018**, *247*, 528–536. [CrossRef] [PubMed]
17. Wijekoon, K.C.; Hai, F.I.; Kang, J.; Price, W.E.; Cath, T.Y.; Nghiem, L.D. Rejection and fate of trace organic compounds (TrOCs) during membrane distillation. *J. Membr. Sci.* **2014**, *453*, 636–642. [CrossRef]
18. Tufail, A.; Alharbi, S.; Alrifai, J.; Ansari, A.; Price, W.E.; Hai, F.I. Combining enzymatic membrane bioreactor and ultraviolet photolysis for enhanced removal of trace organic contaminants: Degradation efficiency and by-products formation. *Process. Saf. Environ. Prot.* **2021**, *145*, 110–119. [CrossRef]
19. Alharbi, S.K.; Price, W.E. Degradation and Fate of Pharmaceutically Active Contaminants by Advanced Oxidation Processes. *Curr. Pollut. Rep.* **2017**, *3*, 268–280. [CrossRef]
20. Tufail, A.; Price, W.E.; Hai, F.I. A critical review on advanced oxidation processes for the removal of trace organic contaminants: A voyage from individual to integrated processes. *Chemosphere* **2020**, *260*, 127460. [CrossRef]
21. Tufail, A.; Price, W.E.; Mohseni, M.; Pramanik, B.K.; Hai, F.I. A critical review of advanced oxidation processes for emerging trace organic contaminant degradation: Mechanisms, factors, degradation products, and effluent toxicity. *J. Water Process. Eng.* **2020**, 101778. [CrossRef]
22. Mozia, S.; Darowna, D.; Przepiórski, J.; Morawski, A.W. Evaluation of Performance of Hybrid Photolysis-DCMD and Photocatalysis-DCMD Systems Utilizing UV-C Radiation for Removal of Diclofenac Sodium Salt from Water. *Pol. J. Chem. Technol.* **2013**, *15*, 51–60. [CrossRef]
23. Khanzada, N.K.; Farid, M.U.; Kharraz, J.A.; Choi, J.; Tang, C.Y.; Nghiem, L.D.; Jang, A.; An, A.K. Removal of organic micropollutants using advanced membrane-based water and wastewater treatment: A review. *J. Membr. Sci.* **2020**, *598*, 117672. [CrossRef]
24. Chapple, A.; Nguyen, L.N.; Hai, F.I.; Dosseto, A.; Rashid, M.H.O.; Oh, S.; Price, W.E.; Nghiem, L.D. Impact of inorganic salts on degradation of bisphenol A and diclofenac by crude extracellular enzyme from *Pleurotus ostreatus*. *Biocatal. Biotransform.* **2019**, *37*, 10–17. [CrossRef]
25. Li, Y.; Qiao, X.; Zhang, Y.-N.; Zhou, C.; Xie, H.; Chen, J. Effects of halide ions on photodegradation of sulfonamide antibiotics: Formation of halogenated intermediates. *Water Res.* **2016**, *102*, 405–412. [CrossRef] [PubMed]
26. Yang, Y.; Lu, X.; Jiang, J.; Ma, J.; Liu, G.; Cao, Y.; Liu, W.; Li, J.; Pang, S.; Kong, X.; et al. Degradation of sulfamethoxazole by UV, UV/H_2O_2 and UV/persulfate (PDS): Formation of oxidation products and effect of bicarbonate. *Water Res.* **2017**, *118*, 196–207. [CrossRef] [PubMed]
27. Ren, D.; Huang, B.; Xiong, D.; He, H.; Meng, X.; Pan, X. Photodegradation of 17α-ethynylestradiol in dissolved humic substances solution: Kinetics, mechanism and estrogenicity variation. *J. Environ. Sci.* **2017**, *54*, 196–205. [CrossRef]
28. Chowdhury, R.R.; Charpentier, P.A.; Ray, M.B. Photodegradation of 17β-estradiol in aquatic solution under solar irradiation: Kinetics and influencing water parameters. *J. Photochem. Photobiol. A Chem.* **2011**, *219*, 67–75. [CrossRef]
29. Chen, Y.; Liu, L.; Su, J.; Liang, J.; Wu, B.; Zuo, J.; Zuo, Y. Role of humic substances in the photodegradation of naproxen under simulated sunlight. *Chemosphere* **2017**, *187*, 261–267. [CrossRef]

30. Asif, M.B.; Fida, Z.; Tufail, A.; Van De Merwe, J.P.; Leusch, F.D.; Pramanik, B.K.; Price, W.E.; Hai, F.I. Persulfate oxidation-assisted membrane distillation process for micropollutant degradation and membrane fouling control. *Sep. Purif. Technol.* **2019**, *222*, 321–331. [CrossRef]
31. Lekkerkerker-Teunissen, K.; Benotti, M.J.; Snyder, S.A.; Van Dijk, H.C. Transformation of atrazine, carbamazepine, diclofenac and sulfamethoxazole by low and medium pressure UV and UV/H2O2 treatment. *Sep. Purif. Technol.* **2012**, *96*, 33–43. [CrossRef]
32. Gago-Ferrero, P.; Badia-Fabregat, M.; Olivares, A.; Piña, B.; Blánquez, P.; Vicent, T.; Caminal, G.; Díaz-Cruz, M.S.; Barceló, D. Evaluation of fungal- and photo-degradation as potential treatments for the removal of sunscreens BP3 and BP1. *Sci. Total Environ.* **2012**, *427–428*, 355–363. [CrossRef] [PubMed]
33. Mozia, S.; Tsumura, T.; Toyoda, M.; Morawski, A.W. Degradation of Ibuprofen Sodium Salt in a Hybrid Photolysis—Membrane Distillation System Utilizing Germicidal UVC Lamp. *J. Adv. Oxid. Technol.* **2011**, *14*, 31. [CrossRef]
34. Mozia, S. Photocatalytic membrane reactors (PMRs) in water and wastewater treatment. A review. *Sep. Purif. Technol.* **2010**, *73*, 71–91. [CrossRef]
35. Hao, Z.; Guo, C.; Lv, J.; Zhang, Y.; Zhang, Y.; Xu, J. Kinetic and mechanistic study of sulfadimidine photodegradation under simulated sunlight irradiation. *Environ. Sci. Eur.* **2019**, *31*, 40. [CrossRef]
36. Koumaki, E.; Mamais, D.; Noutsopoulos, C.; Nika, M.-C.; Bletsou, A.A.; Thomaidis, N.S.; Eftaxias, A.; Stratogianni, G. Degradation of emerging contaminants from water under natural sunlight: The effect of season, pH, humic acids and nitrate and identification of photodegradation by-products. *Chemosphere* **2015**, *138*, 675–681. [CrossRef] [PubMed]
37. Zhang, N.; Liu, G.; Liu, H.; Wang, Y.; He, Z.; Wang, G. Diclofenac photodegradation under simulated sunlight: Effect of different forms of nitrogen and Kinetics. *J. Hazard. Mater.* **2011**, *192*, 411–418. [CrossRef]
38. Liu, H.; Zhao, H.; Quan, X.; Zhang, Y.; Chen, S. Formation of Chlorinated Intermediate from Bisphenol A in Surface Saline Water under Simulated Solar Light Irradiation. *Environ. Sci. Technol.* **2009**, *43*, 7712–7717. [CrossRef]
39. Ribeiro, A.R.L.; Moreira, N.F.; Puma, G.L.; Silva, A.M. Impact of water matrix on the removal of micropollutants by advanced oxidation technologies. *Chem. Eng. J.* **2019**, *363*, 155–173. [CrossRef]
40. Luo, R.; Li, M.; Wang, C.; Zhang, M.; Khan, M.A.N.; Sun, X.; Shen, J.; Han, W.; Wang, L.; Li, J. Singlet oxygen-dominated non-radical oxidation process for efficient degradation of bisphenol A under high salinity condition. *Water Res.* **2019**, *148*, 416–424. [CrossRef]
41. Grebel, J.E.; Pignatello, J.J.; Mitch, W.A. Impact of Halide Ions on Natural Organic Matter-Sensitized Photolysis of 17β-Estradiol in Saline Waters. *Environ. Sci. Technol.* **2012**, *46*, 7128–7134. [CrossRef]
42. Li, F.; Kong, Q.; Chen, P.; Chen, M.; Liu, G.; Lv, W.; Yao, K. Effect of halide ions on the photodegradation of ibuprofen in aqueous environments. *Chemosphere* **2017**, *166*, 412–417. [CrossRef] [PubMed]
43. Lee, Y.-M.; Lee, G.; Kim, M.-K.; Zoh, K.-D. Kinetics and degradation mechanism of Benzophenone-3 in chlorination and UV/chlorination reactions. *Chem. Eng. J.* **2020**, *393*, 124780. [CrossRef]
44. Lee, Y.-M.; Lee, G.; Zoh, K.-D. Benzophenone-3 degradation via UV/H2O2 and UV/persulfate reactions. *J. Hazard. Mater.* **2021**, *403*, 123591. [CrossRef] [PubMed]
45. Mazellier, P.; Busset, C.; Delmont, A.; De Laat, J. A comparison of fenuron degradation by hydroxyl and carbonate radicals in aqueous solution. *Water Res.* **2007**, *41*, 4585–4594. [CrossRef]
46. Merouani, S.; Hamdaoui, O.; Saoudi, F.; Chiha, M.; Pétrier, C. Influence of bicarbonate and carbonate ions on sonochemical degradation of Rhodamine B in aqueous phase. *J. Hazard. Mater.* **2010**, *175*, 593–599. [CrossRef]
47. Huang, W.; Bianco, A.; Brigante, M.; Mailhot, G. UVA-UVB activation of hydrogen peroxide and persulfate for advanced oxidation processes: Efficiency, mechanism and effect of various water constituents. *J. Hazard. Mater.* **2018**, *347*, 279–287. [CrossRef]
48. Yuan, H.; Zhou, X.; Zhang, Y.L. Degradation of Acid Pharmaceuticals in the UV/H 2 O 2 Process: Effects of Humic Acid and Inorganic Salts. *Clean Soil Air Water* **2012**, *41*, 43–50. [CrossRef]
49. Calza, P.; Vione, D.; Minero, C. The role of humic and fulvic acids in the phototransformation of phenolic compounds in seawater. *Sci. Total Environ.* **2014**, *493*, 411–418. [CrossRef]
50. Zhang, Y.; Zhao, F.; Wang, F.; Zhang, Y.; Shi, Q.; Han, X.; Geng, H. Molecular characteristics of leonardite humic acid and the effect of its fractionations on sulfamethoxazole photodegradation. *Chemosphere* **2020**, *246*, 125642. [CrossRef]
51. Wang, Y.; Roddick, F.A.; Fan, L. Direct and indirect photolysis of seven micropollutants in secondary effluent from a wastewater lagoon. *Chemosphere* **2017**, *185*, 297–308. [CrossRef]

52. Chianese, S.; Iovino, P.; Leone, V.; Musmarra, D.; Prisciandaro, M. Photodegradation of Diclofenac Sodium Salt in Water Solution: Effect of HA, NO$_3$—and TiO$_2$ on Photolysis Performance. *Water Air Soil Pollut.* **2017**, *228*, 270. [CrossRef]
53. Trovó, A.G.; Nogueira, R.F.; Agüera, A.; Sirtori, C.; Fernández-Alba, A.R. Photodegradation of sulfamethoxazole in various aqueous media: Persistence, toxicity and photoproducts assessment. *Chemosphere* **2009**, *77*, 1292–1298. [CrossRef] [PubMed]

Publisher's Note: MDPI stays neutral with regard to jurisdictional claims in published maps and institutional affiliations.

© 2020 by the authors. Licensee MDPI, Basel, Switzerland. This article is an open access article distributed under the terms and conditions of the Creative Commons Attribution (CC BY) license (http://creativecommons.org/licenses/by/4.0/).

Article

Evaluating the Feasibility of Forward Osmosis in Diluting RO Concentrate Using Pretreatment Backwash Water

Susanthi Liyanaarachchi [1], Veeriah Jegatheesan [1,*], Li Shu [1], Ho Kyong Shon [2], Shobha Muthukumaran [3] and Chun Qing Li [1]

[1] School of Engineering, RMIT University, Melbourne, VIC 3000, Australia; susanthiliya@gmail.com (S.L.); li.shu846@gmail.com (L.S.); chunqing.li@rmit.edu.au (C.Q.L.)
[2] School of Civil and Environmental Engineering, University of Technology Sydney, Broadway, NSW 2581, Australia; Hokyong.Shon-1@uts.edu.au
[3] College of Engineering & Science, Victoria University, Melbourne, VIC 8001, Australia; Shobha.Muthukumaran@vu.edu.au
* Correspondence: jega.jegatheesan@rmit.edu.au

Received: 7 February 2020; Accepted: 21 February 2020; Published: 25 February 2020

Abstract: Forward osmosis (FO) is an excellent membrane process to dilute seawater (SW) reverse osmosis (RO) concentrate for either to increase the water recovery or for safe disposal. However, the low fluxes through FO membranes as well the biofouling/scaling of FO membranes are bottlenecks of this process requiring larger membrane area and membranes with anti-fouling properties. This study evaluates the performance of hollow fibre and flat sheet membranes with respect to flux and biofouling. Ferric hydroxide sludge was used as impaired water mimicking the backwash water of a filter that is generally employed as pretreatment in a SWRO plant and RO concentrate was used as draw solution for the studies. Synthetic salts are also used as draw solutions to compare the flux produced. The study found that cellulose triacetate (CTA) flat sheet FO membrane produced higher flux (3–6 L m^{-2} h^{-1}) compared to that produced by polyamide (PA) hollow fibre FO membrane (less than 2.5 L m^{-2} h^{-1}) under the same experimental conditions. Therefore, long-term studies conducted on the flat sheet FO membranes showed that fouling due to ferric hydroxide sludge did not allow the water flux to increase more than 3.15 L m^{-2} h^{-1}.

Keywords: biofouling; fertilizers; flat sheet; flux; forward osmosis (FO); hollow fibre; reverse osmosis (RO)

1. Introduction

Diminishing freshwater resources pose a serious threat to various practices. For example, if the water required to make fertiliser solutions can be sourced from impaired water, it will significantly help to conserve freshwater sources for other activities in agricultural farms. Similarly, if the brine produced in a seawater reverse osmosis (SWRO) plant can be diluted using the pre-treatment filter backwash water, it can be reused as the feed to the RO or can be discharged safely to the receiving environment [1,2]. When a Forward Osmosis (FO) membrane separates diluted feed stream (impaired water) and concentrated draw stream (concentrated fertilizer or RO brine solutions), water will naturally pass through the FO membrane from the dilute stream to the concentrated stream to produce diluted solutions. This is due to the osmotic pressure difference created by those two streams. The larger the osmotic pressure difference, the higher the water flux through the membrane. At the same time, reverse salt flux, RSF (movement of salts from the concentrated stream to the dilute stream) would also occur, which can be minimised by selecting appropriate membranes. Generally, FO membranes will have active and support layers which will reduce the effective osmotic pressure difference between the two surfaces of the active layer and thus will reduce the water flux [3–5].

The FO membranes can be obtained either as flat sheets or hollow fibres. Flat sheet membranes available to date are showing lower water fluxes when the concentrations of draw solutions are low [1,6–12]. One of the best available cellulose triacetate (CTA) flat sheet membranes manufactured by Hydation Technology Innovations (HTI), USA, provides a maximum water flux of 9.6 L m^{-2} h^{-1} (LMH) when deionised (DI) water and 0.6 M NaCl salt solution were used as feed and draw solutions, respectively [13].

According to Li et al. [14], CTA FO membranes are made by adding dried CTA and cellulose acetate (CA) polymers to a premixed solvent of dioxane, acetone, lactic acid, and methanol at a certain ratio. The polymer/solvent solution will be kept at 30 °C and stirred till a homogeneous solution is obtained. The solution will then be stored in an oven at 30 °C for several hours to de-aerate and then will be cast onto a dry clean glass plate. The formed film will be immersed into a water bath within 3 s at 12 °C. After solidification, the membranes will be immersed in deionized water for 24 h before conducting any tests.

According to Lim et al. [15], a typical dry-jet wet spinning method can be applied for preparation of the hollow fibre membrane substrates. Dried polyether sulfone (PES) powder and polyethylene glycol (PEG400) at a fixed amount can be mixed with N-Methyl-2-pyrrolidone (NMP) at 60 °C for 12 h. Hydrophilic non-solvent (PEG) is added into the polymer solution to produce a sponge-like porous morphology for enhancing pore formation and interconnection. A degassed polymer solution will then be pumped into the double spinneret nozzle together with the bore fluid and the molded fibres will be immersed into the coagulation bath immediately. The solidified substrates will then be rolled up and stored in DI water for 24 h. The hollow fibre membranes will be immersed in the aqueous glycerol solution (50 wt%) for two days and dried in the atmosphere to minimise the collapse of their pore structures in open-air storage. Hollow fibre membrane modules can be made using the fibres [15].

This study evaluates the performance of a hollow fibre membrane over a flat sheet membrane with respect to the water flux. Based on the flux results, biofouling of flat sheet membranes was also evaluated.

2. Materials and Methods

2.1. Membranes

Flat sheet cellulose tri-acetate (CTA) membranes were purchased from HTI, USA. The support layer of the flat sheet membrane is made up of polyester mesh and average pore diameter is 0.74 nm [16]. Scanning Electron Microscopy (SEM) images of the flat sheet CTA membrane are given in Figure 1a–c. As Figure 1a shows [17], the membrane is on an embedded screen support. Figure 1b shows the support layer and the embedded mesh and Figure 1c is the active layer where water permeation happens.

Hollow fibre polyamide (PA) membranes used were fabricated at Samsung Cheil Industries Inc., South Korea and consist of a Sulphonated Polysulphone (SPSf) support layer. SEM images of the hollow fibre PA membrane are given in Figure 1d,e [18]. Figure 1d shows the thickness of the lumens with pores, and Figure 1e shows the lumens. CTA and PA membranes used were hydrophilic and hydrophobic, respectively.

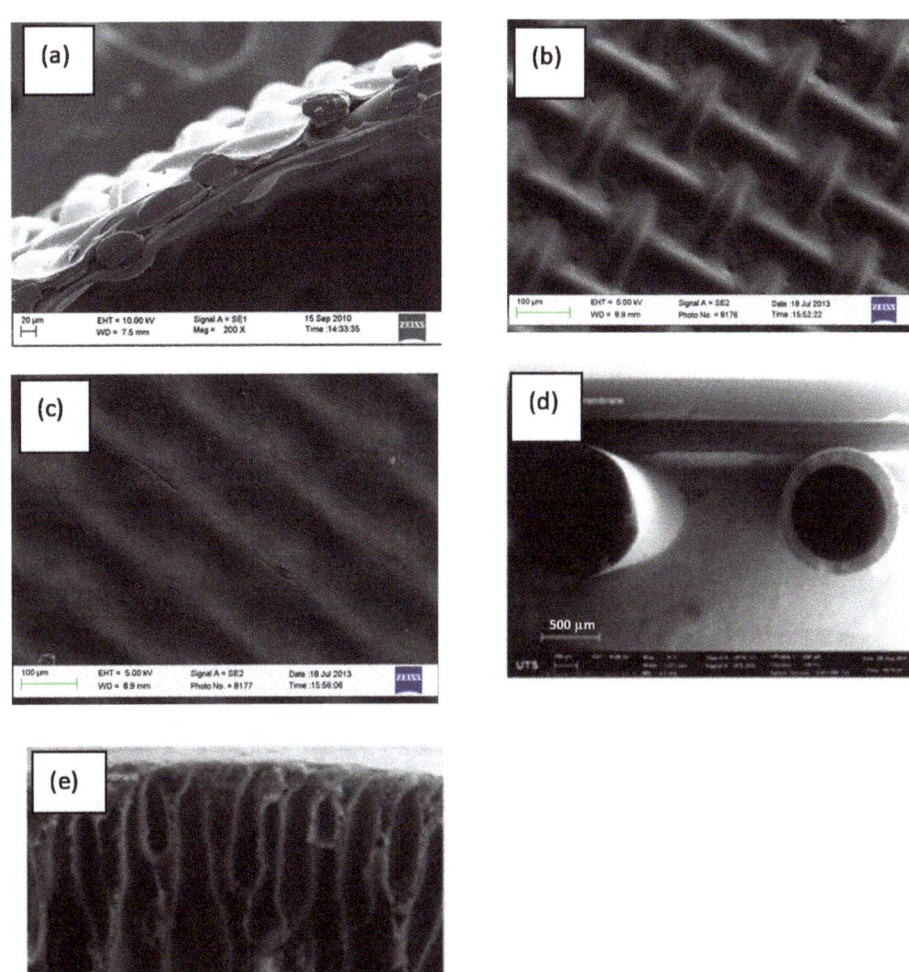

Figure 1. Images of hydrophilic Cellulose Triacetate (CTA) membrane on embedded polyester screen support: (**a**) cross section (reprinted from [17] with permission from Elsevier); (**b**) Support side; (**c**) active side. SEM images of hydrophobic Polyamide (PA) hollow fibre membrane (reprinted from [18] with permission from Elsevier) with inner surface of the hollow fibre as active layer, and the outer surface as support layer; (**d**) cross section showing the thickness of the lumens along with pores (**e**) enlarged cross section showing the lumen.

2.2. Flux Studies with Hollow Fibre Membrane

Feed (either deionised (DI) water or $Fe(OH)_3$ sludge representing impaired water) and draw solutions ($NaCl$, $MgCl_2$, $CaCl_2$, Na_2SO_4 and seawater reverse osmosis concentrate (ROC) representing concentrated solution) were passed through the polyamide (PA) hollow fibre FO membrane at different feed and draw Reynolds number (*Re*) ratios. Reynolds numbers were varied by changing the velocity of the feed and draw solutions. Sludge/DI water was circulated outside the hollow fibre membrane and the draw solution through the lumen side. Since the lumen side surface of the hollow fibre is the active layer, the experiments have been run at an active layer facing draw solution (AL-DS) mode. Even though the sludge particles may block the support side in this mode, cleaning the outer surface of the fouled hollow fibre membranes will be much easier compared to cleaning the inner lumen side of the membrane. The experimental set up is shown in Figure 2. Change in the weight of the draw solution was programmed to be stored in a data logger at one-minute time intervals. Experimental water flux ($J_{w,e}$) was determined by the following equation:

$$J_{w,e} = \frac{\text{change in weight of the draw solution in time } \Delta t}{\text{density of water } \times \text{effective membrane area } \times \Delta t} \tag{1}$$

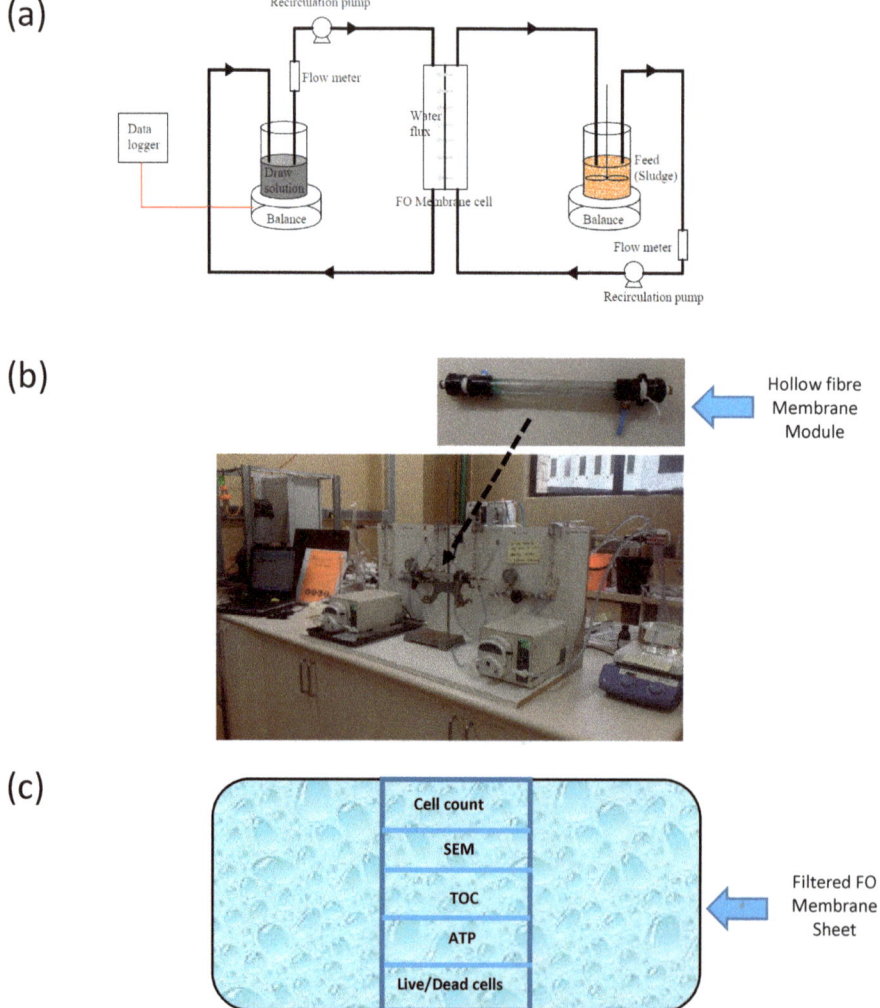

Figure 2. Hollow fibre membrane (**a**) schematic of the experimental set up with a general forward osmosis (FO) membrane cell that can accommodate either a flat sheet or a hollow fibre membrane module, (**b**) photo of the hollow fibre FO membrane system used in this study (Effective membrane area is 25.45 cm^2), and (**c**) protocol for the analyses of fouled membrane.

After 1 h of filtration, properties of the feed and draw solutions were measured. The membrane was cleaned using DI water prior to each experiment.

2.3. Fouling Studies with Flat Sheet Membranes

The draw and feed solutions were ROC and pre-treatment sludge (filter backwash water) from a seawater RO desalination plant, respectively. Thus, the solids content of pre-treatment sludge was varied from 2% to 8% of total solids (TS). This solids content represents the suspended solids that are removed from the filter during the backwash. This is because there are two types of pre-treatment sludge that can be generated in an RO desalination plant; the media filters used for the pre-treatment of seawater can be backwashed using either pre-treated seawater or ROC and can produce pre-treatment sludge with different total solids and ionic strength. However, the dewatered sludge available in the lab had 15% TS as received from Perth Seawater Desalination Plant (PSDP). Therefore, to obtain required TS contents of each pre-treatment sludge, 15% TS sludge was diluted using (i) pre-treated seawater (and the feed solution obtained was named as High EC with an EC of 45 mS/cm) and (ii) DI water (and the feed solution obtained was named as low EC with an EC of 1.5 mS/cm). Prepared feed and draw solutions were passed through the cellulose triacetate (CTA) flat sheet FO membrane at 0.04 m/s cross flow velocity in counter current flow configuration [19,20].

Pre-treatment sludge was circulated on the support side of the membrane (FO mode) and stirred at a constant rate during the experiment to eliminate settling of particles. Experimental set up was similar to that in Figure 2a. Photos of the hollow fibre membrane module as well as the experimental setup are shown in Figure 2b. Experiments were run at 20 ± 2 °C and triplicated at each operating condition. Again, change in the weight of the draw solution was programmed to be stored in a data logger at 5 min time intervals. Furthermore, three consecutive experimental setups (similar to Figure 2a) were run. Fouling behaviour on the FO membrane was examined after one day, four days, one week and five weeks. One experiment was run until the membrane was fully fouled (i.e., until no water flux observed). Water flux, conductivity, total organic carbon (TOC), and pH of each set up were monitored continuously using a data logger, electrical conductivity (EC) meter, TOC analyser, and pH meter, respectively.

All of the fouling experiments were run in semi-batch mode as the experiments were long-term runs, following the experimental procedure of Li et al. [21], i.e., when the draw solution had extracted 15% of water from the feed (150 mL), both draw and feed solutions were replaced with fresh 1L tank; TOC, pH, temperature, and EC of the replaced solutions were measured. Prior to each new experiment, three experimental setups were thoroughly cleaned to remove trace organic matter using the procedure given in the next section [22]. It is important to note that the short-time treatment with alkaline hypochlorite solution could improve the membrane performance slightly [23]. Accordingly, the hypochlorite degradation reaction of aromatic PA membrane involves a reversible and an irreversible chlorination. The reversible intermediate could be regenerated to initial amide by the treatment with alkaline before it rearranged to an irreversible product, thus partially improving the membrane performance.

2.4. Cleaning of FO Set-up to Remove Trace Organic Impurities Prior to Each Fouling Test

The following procedure was followed to clean the FO set-up:

I. Recirculation of 0.5% sodium hypochlorite through the FO set-up for 2 h.
II. Removal of trace organic matter by recirculating 5 mM ethylene di-amine tetra-acetic acid (EDTA) at pH 11 through the set-up for 30 min.
III. Additional removal of trace organic matter by recirculating 2 mM sodium dodecyl sulphate (SDS) at pH 11 through the set-up for 30 min.
IV. Sterilisation of the unit by recirculating 95% ethanol through the set-up for 1 h.
V. Rinsing the unit with DI water (several times) to eliminate ethanol residue.

Once the filtration was complete, a known area of membrane was selected for analysis for cell count, SEM, TOC, adenosine triphosphate (ATP), and live/dead cell count analysis (Figure 2c). Membrane fouling is due to the deposition of chemical (organic/inorganic) species and biological growth on the membrane surface. Thus, cell count is an indicator of biological growth and hence the progress of biofouling of the membrane.

3. Results

The properties of synthetic draw solutions used in this study are given in Table 1. Out of the five synthetic draw solutions used (1.0 M NaCl, 1.0 M $MgCl_2$, 1.0 M $CaCl_2$, 1.0 M Na_2SO_4 and ROC), Na_2SO_4 has the largest density while $MgCl_2$ has the largest viscosity; $CaCl_2$ has the highest electrical conductivity. Properties of seawater (as feed) and reverse osmosis concentrate (ROC) as drawn are given in Table 2.

Table 1. Properties of synthetic draw solutions used in this study.

Draw Solution (1 M)	Density, ρ (kg/m^3)	Viscosity, μ (Pa·s)	Conductivity *, EC (mS/cm)	Osmotic Pressure at 25 °C (bar)
NaCl	1037.00	0.001080	81.1	46.4
Na_2SO_4	1557.00	0.001120	81.9	52.0
$MgCl_2$	1072.40	0.001490	96.7	79.9
$CaCl_2$	1085.20	0.001330	108.6	80.0
ROC	1023.98	0.001004	72.3	33.0

Note: Density, viscosity and osmotic pressure were obtained from the OLI ® stream analyser, and * conductivity from experimental values.

Table 2. Properties of feed and draw solution used in this study.

Property	Seawater	Sand Filtered Seawater	Draw Solution—ROC	Feed Solution—PSDP Fe(OH)$_3$ Sludge
pH	8.42	7.68	7.77	8.69
Turbidity (NTU)	29.1	0.45	-	-
EC (mS/m)	4450	4470	7300	5150
TOC (mg/L)	1.71	0.73	3.10	17.06
Alkalinity—mg/L as CaCO$_3$	110	45	68	102
Hardness (EDTA)-mg/L as CaCO$_3$	4600	6200	9550	4500
Solids content (% TS)	-	-	-	4.04
Specific gravity	-	-	-	1.01

3.1. Effect of Re on the Water Flux

Figure 3 shows the water flux through hollow fibre FO membranes when DI water and salt solutions were used as feed and draw solutions, respectively. Draw Re was kept at 1000 and 2000 while feed Re was kept at 200, 450, and 1200. Thus, six experiments with different feed and draw Re were conducted. Water flux of up to 10 LMH was observed when a laminar condition (Re = 1000) prevailed in the flow of draw solution. The Na_2SO_4 draw solution gave the highest flux similar to that of $MgCl_2$. This is interesting as 1.0 M $MgCl_2$ has the highest osmotic pressure; however, when the Re is similar, 1.0 M Na_2SO_4 shows similar performance even though its osmotic pressure is lower. Furthermore, when the Re of draw solution flow was increased to become near turbulent (at Re = 2000), all three draw solutions showed better performance compared to $MgCl_2$. The $MgCl_2$ solution drew a maximum of 5.1 LMH when the feed Re was the highest (1200). Therefore, it is evident that, when selecting a draw solution, not only its osmotic pressure, but also its viscosity, density, and the crossflow velocity are affecting the performance in terms of water flux.

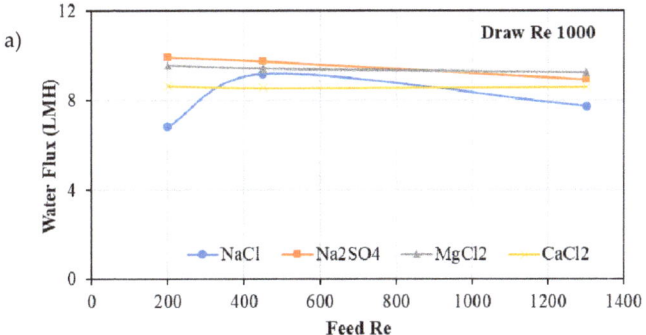

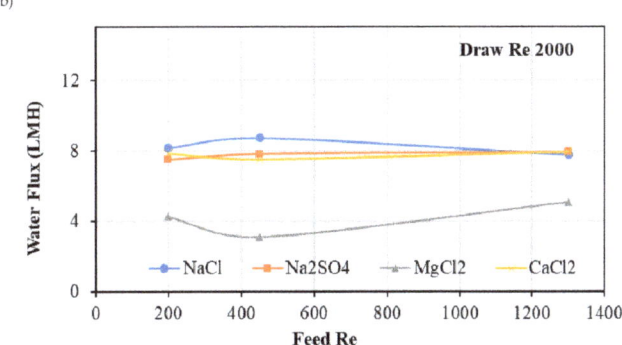

Figure 3. Flux through hollow fibre membranes when draw solution Re was (**a**) 1000 and (**b**) 2000. Note that the experiments were run in active layer facing draw solution (AL-DS) mode to compare the results with sludge dewatering experiments.

Despite the type of solution, the feed flow with a *Re* of 200 and the draw solution flow with a *Re* of 1000 gave the best performance in terms of water flux. Low *Re* of feed flow provide enough time for the water to pass through the membrane and high *Re* of the draw flow reduces the dilution of the draw solution as it takes away the water flux coming from the feed side of the membrane quicker. This allows the FO process to produce high flux under those conditions. Therefore, sludge dewatering experiments (detailed in the following section) were conducted at 200:1000 feed to draw an *Re* ratio.

Reverse salt flux of the membrane was determined by measuring the EC values of the feed solution. Since the feed solution was DI water, the change in EC was obviously due to the ions transported through the membrane from the draw solution. Figure 4 shows the RSF (or EC values of the feed solution) for each draw solution. NaCl shows the highest increase in RSF with time. Despite the *Re*, RSF is increasing with the filtration time. $CaCl_2$ shows the lowest RSF (below 5 μS/cm). $MgCl_2$ which showed lower water fluxes compared to the other salt solutions shows lower RSF—however, higher than that of $CaCl_2$. In general, the RSF in hollow fibre membrane was found to be small which was also the case in the literature [24,25]. Addition of divalent ions into the NaCl draw solution reduced the RSF but did not affect the flux, and $MgCl_2$ was found to be a better additive [18]. However, multi-valent cations can form complexes with organic and colloidal foulants and expedite fouling and cake-enhanced osmotic pressure [26–29].

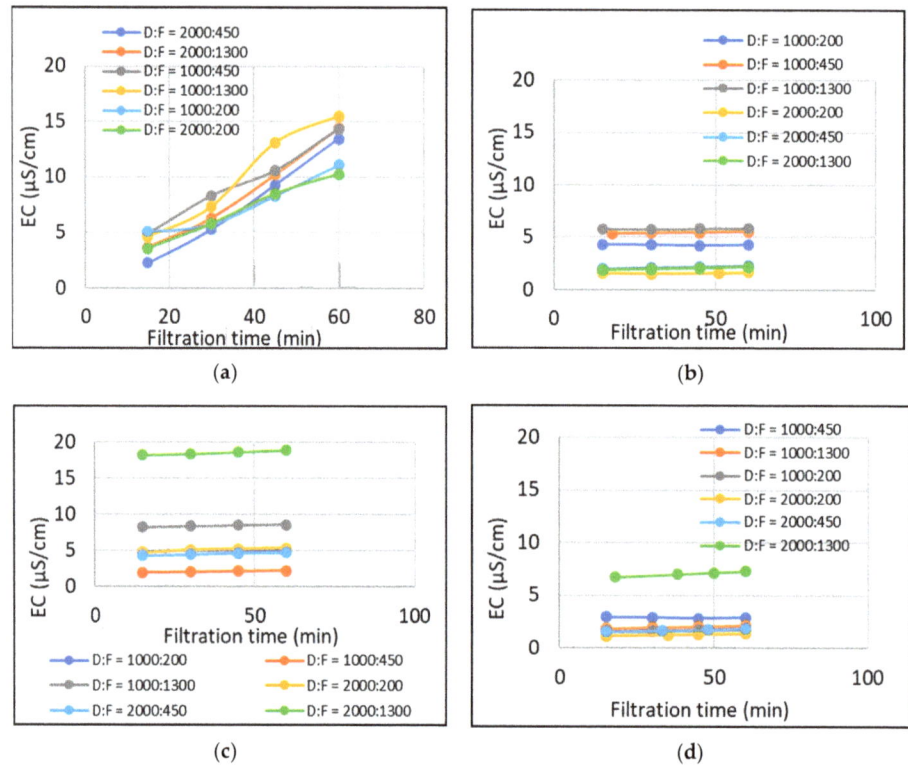

Figure 4. Measurements during filtration in hollow fibre FO membrane with different draw solutions (**a**) 1.0 M NaCl; (**b**) 1.0 M $CaCl_2$; (**c**) 1.0 M $MgCl_2$; (**d**) 1.0 M Na_2SO_4.

3.2. Effect of Sludge Solids Content

Figure 5 shows the amount of dewatered sludge from PSDP required to produce pre-treatment sludge with different total solids concentration. Since seawater and DI water were used for dilution, the vertical gap between the two graphs should be the TS content of the seawater. Therefore, the vertical gap 3.374 (= 3.3964 − 0.0224) should be the TS% in the seawater used to prepare pre-treatment sludge. Since TDS of seawater is 30–35 g/L, 3.374 TS% appears acceptable. For FO dewatering applications through hollow fibre membranes, low EC sludge samples were chosen assuming lower EC (hence, higher EC difference between the feed and the draw solution) would give better performance with the membrane.

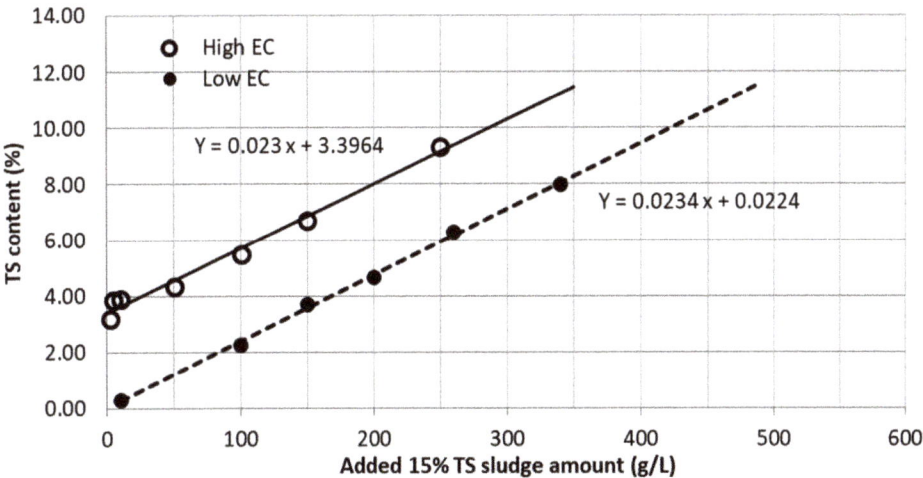

Figure 5. TS content of sludge with high and low EC (prepared in the laboratory starting with dewatered sludge having 15% TS obtained from PSDP).

Since the same membrane was used for each experiment (after cleaning), before and after the two-hour sludge dewatering, baseline experiments were run with 0.5 M NaCl and DI water as draw and feed solutions, respectively. This was to check whether the membrane coupon had returned to the initial condition after cleaning. The results are shown in Figure 6. As Figure 6 illustrates, cleaning has taken the membrane back to the original condition. This means, since the sludge dewatering time was only two hours, the membrane was either not fouled or the fouling is nearly 100% reversible. However, to compare the water flux at each sludge solids content, averaged water fluxes were plotted in one graph as shown in Figure 7. As Figure 7 shows, the lowest sludge solids content led to the highest water flux, i.e., 3.6 LMH, whereas all the other sludge types showed a flux of 1.5–2.5 LMH. When sludge solids content increased, there was a slight drop in the water flux. With increase in solids content, the viscosity and the density of the sludge increase. Higher viscosity means lower Re, and higher density means higher Re; however, the combination of higher viscosity and higher density led to lower water permeation through the hollow fibre membranes. The effect of higher amount of solids content was dominant, and this would have increased the concentration polarisation (CP) effect as sludge passed through the porous side of the membrane leading to lower water flux.

In our previous study [30], we conducted flux experiments with 4.04% pre-treatment sludge as feed solution and ROC as draw solution. CTA flat sheet FO membrane (from HTI USA) was used in the study as well. The active layer facing feed solution (AL-DS) mode gave a water flux of around 3 L m^{-2} h^{-1}, which is higher than the flux obtained under the same condition in this study using hollow fibre membranes (flux between 2 and 2.5 when the % TS sludge varied from 3.68 to 4.67 as shown in Figure 7). Therefore, we decided to use flat sheet membranes for further studies on fouling.

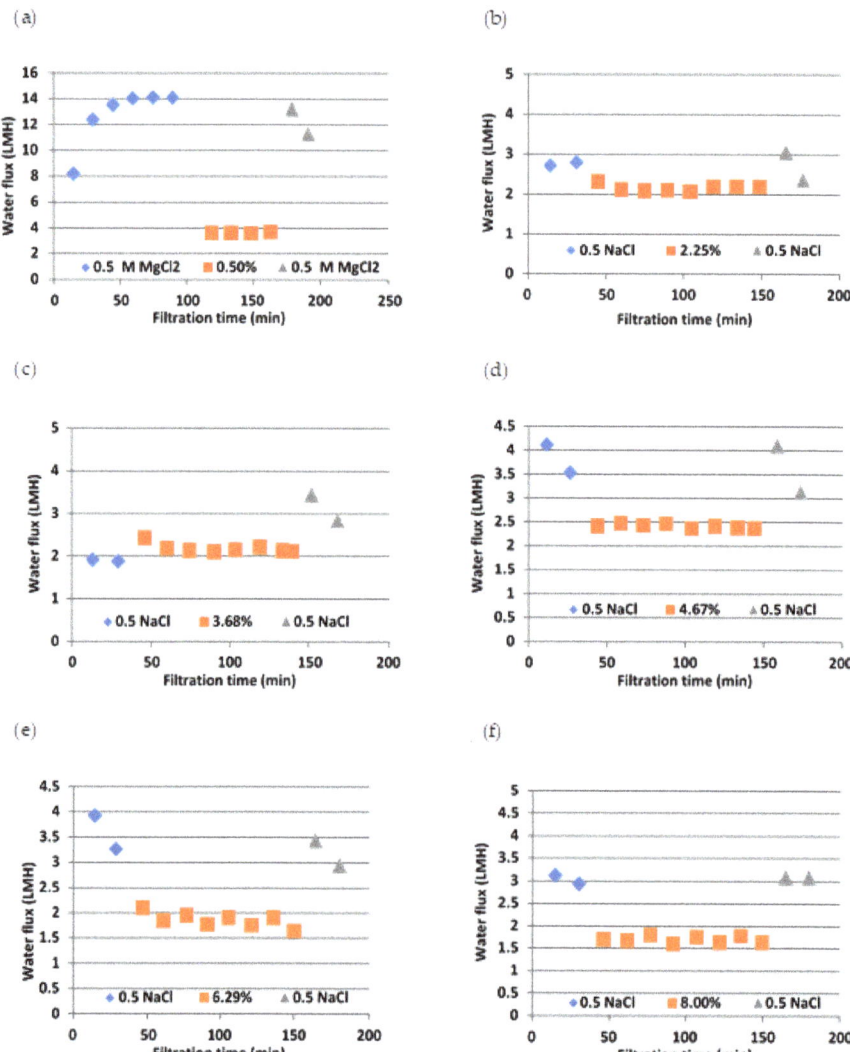

Figure 6. Flux at different sludge contents. Experiments were conducted at first with either MgCl$_2$ (**a**) or NaCl (**b–f**) as draw solutions and DI as feed solution (shown in ♦) and then different concentrations of sludge were used as feed solution and reverse osmosis concentrate (ROC) as draw solution (shown in ■), and, finally, the membrane was cleaned and experiments were reverted back to original MgCl$_2$ or NaCl as draw solutions and DI as feed solution (shown in ▲).

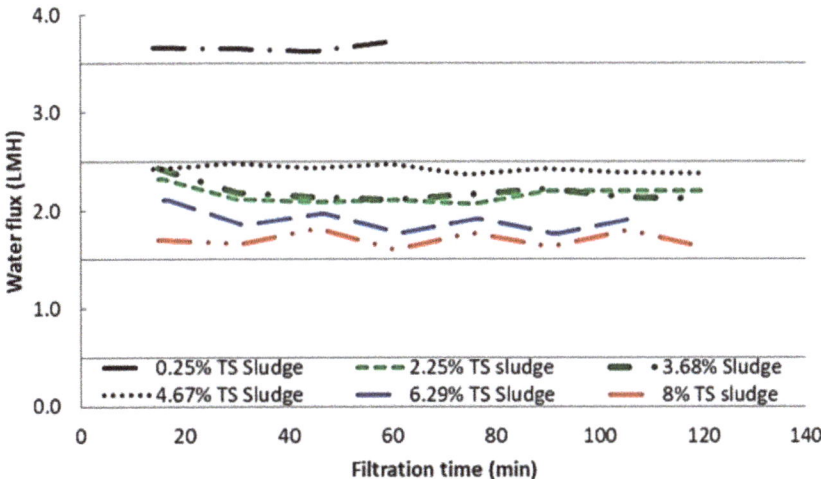

Figure 7. Comparison of water flux at each sludge solids concentrations.

3.3. Fouling Studies with Flat Sheet Membranes

Fouling of FO membrane is an important factor which will hinder the effectiveness of the process if not understood well and controlled through appropriate mitigative steps. However, sometimes, fouling will be beneficial depending on the application of the FO process. For example, Valladares Linares et al. [31] studied the effect of fouling of FO membrane in the rejection of micro-pollutants. They demonstrated clearly that fouled FO membrane rejected the hydrophilic ionic compounds and hydrophobic neutral compounds highly. Rejection of hydrophilic neutral compounds reduced slightly. They concluded that using FO along with low-pressure RO system can provide a barrier to the passage of micro-pollutants present in the secondary wastewater effluent. Such findings on treating wastewaters with high fouling propensities have been treated by FO processes are reported in other studies as well [32,33]. In another study, both reversible and irreversible membrane fouling were absent during the FO experiments when the active layer of the FO membrane was facing the activated sludge solution [34]. One study noted that the flux was strongly dependent on the type of FO membrane used, implying the varying level of fouling on different membranes [35]. Osmotic membrane bioreactors [36,37] are used to improve the removal of organic and ammonium ions from wastewater. Fouling was controlled by osmotic backwash. Large amounts of nutrients, total and suspended solids in the centrate that emerge from a centrifuge that dewaters digested sludge in a wastewater treatment plant, generally returned back to the inlet of the treatment plant. This stream contributes significantly to the nitrogen and phosphorus load of the influent. Recovering the nutrients using an FO/RO system has been evaluated in a study [36], which indicated that CTA-FO membrane showed less fouling (cake layer compaction) compared to CTA-RO membrane due to the lack of applied pressure. Flux recovery after cleaning of the foulants was also high since it was easy to remove the foulants deposited on the membrane. Furthermore, recent advances in the modification of FO membranes through nano-modifiers have improved anti-fouling properties of the membranes significantly and an excellent review on this by Sun et al. [38] can be found in the literature. Discussions on nano-modifiers such as low dimensional carbon-based nanomaterials (carbon nanotubes, graphene oxides, and carbon nanofibers), other nanomaterials (halloysite nanotubes, boehmite, silica, zeolite, nano-$CaCO_3$), and metal/metal-oxide nanoparticles (silver and TiO_2) can be found in the above-mentioned review [38]. Our study evaluates the performance of unique feed solution on the fouling of FO membrane which can contribute valuable information to the desalination industry.

3.3.1. Change in Water Flux

The water flux pattern with time is shown in Figure 8. Flux declined with filtration time due to two reasons (1) fouling and (2) dilution of the draw solution as draw solution was recirculated. However, flux increased when the draw and feed solutions were replaced with fresh solution. This increased flux was lower than the initial flux of the previous batch due to fouling of the membrane. Flux decline due to fouling is shown in red dashed lines in Figure 8c,d. After one week of filtration, the flux declined further in Figure 8c due to the thickened fouling layer deposited on the membrane. The layer may have contained microorganisms and salt deposits as both draw and feed solutions contain salt ions. However, as the EDX spectrums shown in Figure 9a(iii),b(iii), after one week of continuous filtration, the FO system showed only salt deposits. This fouling could easily be removed by providing regular flushes at high cross flow velocities as deposited layers were thin and loose.

Interestingly, as Figure 8d shows, after 300 h (about 2 weeks), the flux was increased once more; however, it was less than the initial water flux. This was repeated after about 650 h as well (around 4 weeks). After about 2 weeks, the loose salt deposit layer had formed and when its thickness increased, part of the loose layer could be readily removed by the increased cross-flow velocity (because when thickness reduces, velocity increases, as shown in Figure 9c–e).

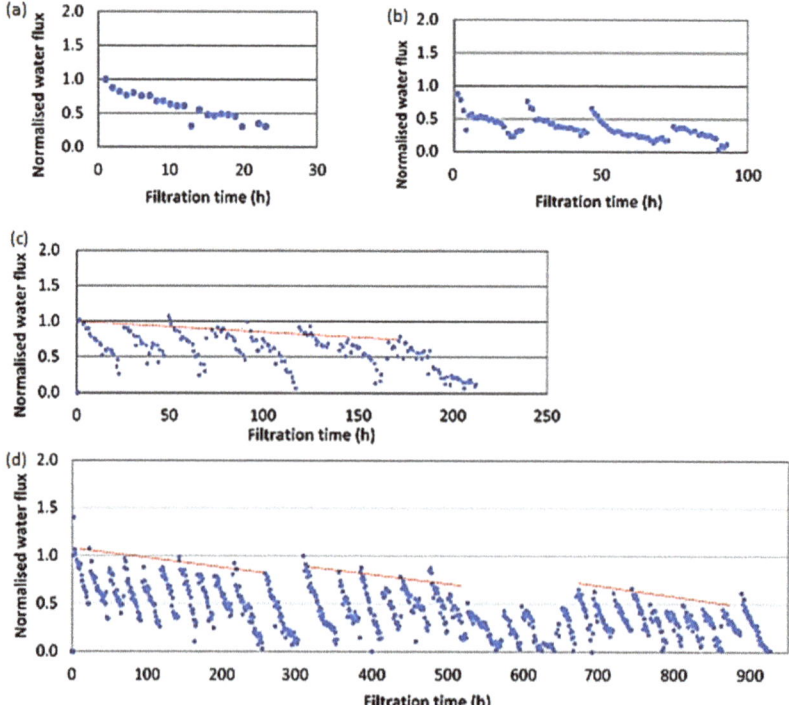

Figure 8. Flux through FO membrane during long-term filtration (**a**) 1 day; (**b**) 4 days; (**c**) 1 week; and (**d**) 5 weeks.

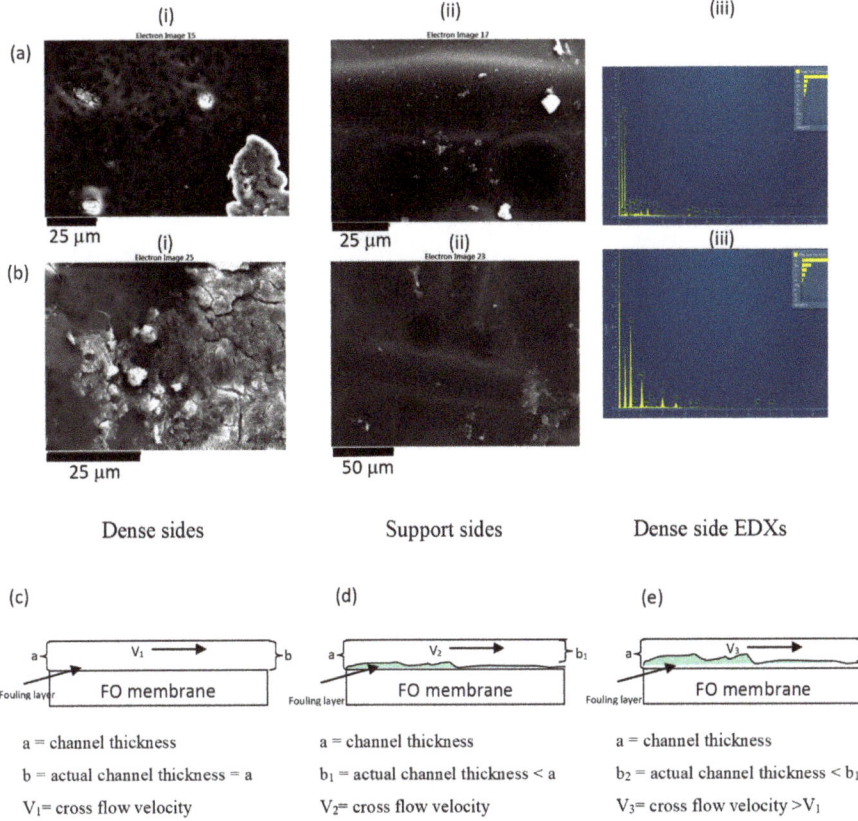

Figure 9. SEM images and EDX spectra of the membrane surface after (**a**) one week, (**b**) five weeks of filtration; schematic diagram of the FO membrane surface with filtration time: (**c**) initially, (**d**) during the first two weeks of operation, and (**e**) after about two weeks of operation. Fouling layer thickness increased over time so does the cross-flow velocity.

In summary, around 50% reduction in water flux was observed due to fouling during five weeks of continuous filtration, without cleaning in between. This is mainly due to deposition of metals. After eight weeks of filtration, there was no water permeation (Figure 10a). Salt deposition on the FO membrane coupon filtered for eight weeks was higher compared to the FO membrane coupon filtered for five weeks (Figure 10b,c). With frequent cleaning with water, water flux can be brought back to initial value as fouling in FO membrane is reversible. A once a week cleaning cycle may be required for longer runs as live (and dead) cells were observed after one week of filtration on the membrane surface.

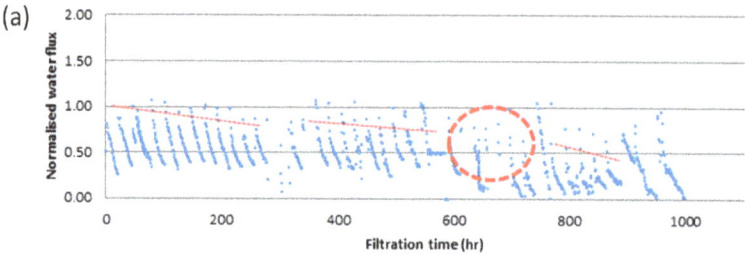

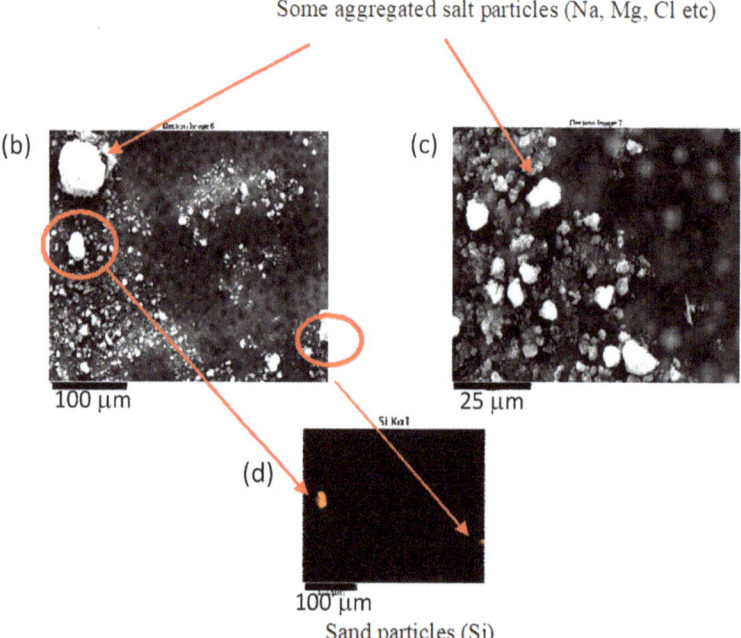

Figure 10. (a) normalised water flux with respect to filtration time; SEM images of the fouled membrane; (b) feed side; (c) draw side; (d) elemental analysis corresponding to (a) obtained through EDX.

3.3.2. Total Organic Carbon (TOC) Results

Draw and feed solutions were replaced with fresh draw and feed samples every 24 h. The used feed and draw solution TOC values were measured daily. Eight weeks of TOC results are reported in Figure 11a. During eight weeks of filtration, TOC of the feed and draw solutions fluctuated. Once the 1-day, 4-day, 1-week, 5-week and 8-week filtration runs were completed, a known area of membrane was selected (as shown in Figure 2c) and vortexed with DI water to extract the deposited fouling layer. The extracted liquid was used to analyse the TOC content per unit of membrane area (Figure 11b). TOC on the membrane surface has increased 10 mg/cm^2 when the filtration time increased from one day to five weeks. In addition, microorganisms started to grow on the membrane surface after one week of continuous filtration. As shown in Figure 11b, live and dead cells were propagated over the membrane surface which then led to reduction of water flux. Therefore, membrane may be needing at least a weekly cleaning cycle to avoid this. In the eight weeks of the filtration trial, the TOC value was significantly low (only 10 mg/cm^2), which is hard to explain why. All the experiments other than an eight-week filtration trail were triplicated. Therefore, another duplicate experiment for an 8-week trial would be required to confirm the TOC results.

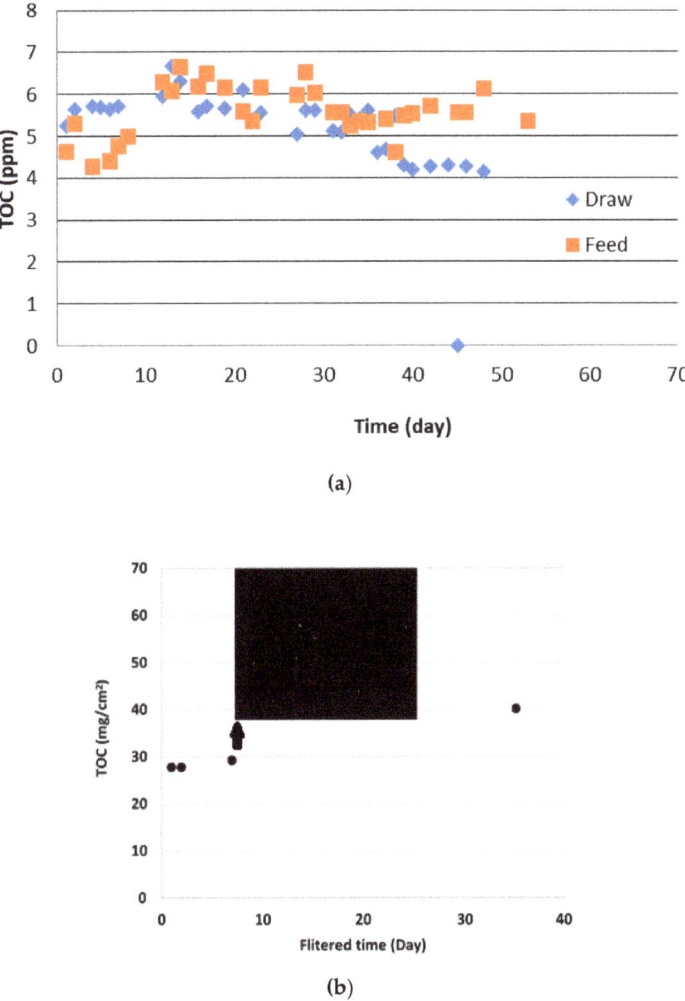

Figure 11. (a) Daily TOC results of the feed and draw solution; (b) TOC of the filtered membrane and live and dead cells on the membrane (Green—live cells and Red—dead cells).

3.4. Future Needs to Improve the Performance of FO Membranes

In order to improve the water flux in a FO process, the internal concentration polarisation (ICP) of solutes should be reduced [39]. ICP is due to the migration of solutes through the porous support layer to the interface between the active and support layers of the membrane. This will reduce the effective osmotic pressure difference between the draw and the feed solution and reduce the flux. One of the ways to reduce ICP is to reduce the structural parameter of the membrane, S defined by the following formula:

$$S = KD = t\tau/\varepsilon \quad (2)$$

where K is the resistance to solute diffusion, D is the diffusion coefficient of the solute; t is the thickness, τ is tortuosity and ε is porosity of the support layer of the membrane. The smaller the S, the larger the water flux. Recently, Shan et al. [40] have developed FO membranes by linking graphene oxide (GO) with oxidized carbon nanotubes (OCNTs) through oxygen-containing groups to form water channels in the polyamide layer. Such membrane had a very low S (230 μm) and showed very high water flux. For such membrane, when 0.5 M NaCl and deionized water were used as draw solution and feed solution, respectively, the water fluxes were 114 and 84.6 LMH for AL-DS and active layer facing feed solution (AL-FS) modes, respectively. The salt fluxes were 5.17 and 3.4 gMH for AL-DS and AL-FS modes, respectively. Further improvement to such membrane was carried out by Kang et al. [41] by assembling graphene oxide (GO) and oxidized carbon nanotubes (OCNTs) with five bilayers on a polyethersulfone membrane to reduce the salt passage through the membrane. An excellent review on membranes similar to the above producing high water fluxes and low reverse salt fluxes is documented by Sun et al. [38].

4. Conclusions

Water fluxes produced by CTA flat sheet and PA hollow fibre FO membranes were compared to select appropriate membrane for further studies on fouling. The Reynolds Number (*Re*) of draw and feed solutions was varied to enhance the water flux through membrane. Lower *Re* of feed and draw solution flows produced better water flux compared to higher *Re* of feed and draw solution flows. When both membranes are used to derive water flux with pre-treatment sludge (or filter backwash water) as feed solution and ROC as draw solution, the PA hollow fibre membrane yielded an average water flux of 2.1 LMH. The process was operated under AL-DS mode, and the sludge solids content in the pretreatment sludge was 3.68%. In our previous study, under similar conditions, flat sheet CTA membranes showed 1.5 times higher water flux compared to PA hollow fibre membranes. Further studies on fouling using the CTA flat sheet membrane confirmed that water flux can decrease by 50% over a period of five weeks due to fouling, if the membrane is not cleaned in between. If the FO process is continued to run without further cleaning, the flux ceases after eight weeks from the beginning of the run. With frequent cleaning with water, water flux can be brought back to the initial value as fouling in the FO membrane is reversible. A once a week cleaning cycle may be required for longer runs and to prevent biofouling. The flux obtained through both PA hollow fibre and CTA flat sheet membranes is less than or equal to 3.15 LMH, which is not sufficient for the system to be economical. Improving the performance of FO membranes by reducing the structural parameter through the introduction of nanomodifiers to the membrane material is one of the ways to go. This will enhance the utilization of FO in places where freshwater resources are diminishing and therefore reusing concentrates and reducing their impacts on fresh water sources are of paramount importance.

Author Contributions: Conceptualization, V.J., L.S. and H.K.S.; methodology, S.L., V.J., L.S., and H.K.S.; software, H.K.S., V.J., and S.L.; validation, S.L., V.J., L.S., and S.M.; formal analysis, S.L., V.J., and S.M.; investigation, V.J., L.S., S.L., and H.K.S.; resources, C.Q.L., V.J., and S.M.; data curation, S.L.; writing—original draft preparation, S.L.; writing—review and editing, V.J.; supervision, V.J., L.S., S.M., and H.K.S.; project administration, C.Q.L. and V.J.; funding acquisition, C.Q.L. and V.J. All authors have read and agreed to the published version of the manuscript.

Funding: S.L. acknowledges the funding support provided by Deakin University, RMIT University, and Victoria University during her PhD study.

Conflicts of Interest: The authors declare no conflict of interest.

References

1. Phuntsho, S.; Shon, H.K.; Hong, S.; Lee, S.; Vigneswaran, S. A novel low energy fertilizer driven forward osmosis desalination for direct fertigation: Evaluating the performance of fertilizer draw solutions. *J. Membr. Sci.* **2011**, *375*, 172–181. [CrossRef]
2. Liyanaarachchi, S.; Jegatheesan, V.; Muthukumaran, S.; Gray, S.; Shu, L. Mass balance for a novel RO/FO hybrid system in seawater desalination. *J. Membr. Sci.* **2016**, *501*, 199–208. [CrossRef]

3. Cath, T.Y.; Childress, A.E.; Elimelech, M. Forward osmosis: Principles, applications, and recent developments. *J. Membr. Sci.* **2006**, *281*, 70–87. [CrossRef]
4. Touati, K.; Tadeo, F. Study of the reverse salt diffusion in pressure retarded osmosis: Influence on concentration polarization and effect of the operating conditions. *Desalination* **2016**, *389*, 171–186. [CrossRef]
5. Ray, S.S.; Chen, S.S.; Nguyen, N.C.; Nguyen, H.T.; Dan, N.P.; Thanh, B.X.; Trang, T. Exploration of polyelectrolyte incorporated with Triton-X 114 surfactant based osmotic agent for forward osmosis desalination. *J. Environ. Manag.* **2018**, *209*, 346–353. [CrossRef]
6. Achilli, A.; Cath, T.Y.; Childress, A.E. Selection of inorganic-based draw solutions for forward osmosis applications. *J. Membr. Sci.* **2010**, *364*, 233–241. [CrossRef]
7. Alejo, T.; Arruebo, M.; Carcelen, V.; Monsalvo, V.M.; Sebastian, V. Advances in draw solutes for forward osmosis: Hybrid organic-inorganic nanoparticles and conventional solutes. *Chem. Eng. J.* **2017**, *309*, 738–752. [CrossRef]
8. Ge, Q.; Ling, M.; Chung, T.-S. Draw solutions for forward osmosis processes: Developments, challenges, and prospects for the future. *J. Membr. Sci.* **2013**, *442*, 225–237. [CrossRef]
9. Ge, Q.; Su, J.; Chung, T.-S.; Amy, G. Hydrophilic Superparamagnetic Nanoparticles: Synthesis, Characterization, and Performance in Forward Osmosis Processes. *Ind. Eng. Chem. Res.* **2011**, *50*, 382–388. [CrossRef]
10. Mathew, R.; Paduano, L.; Albright, J.G.; Miller, D.G.; Rard, J.A. Isothermal Diffusion Coefficients for NaCl-MgCl$_2$-H$_2$O at 25 °C. 3. Low MgCl$_2$ Concentrations with a Wide Range of NaCl Concentrations. *J. Phys. Chem.* **1989**, *93*, 4370–4374. [CrossRef]
11. Su, J.; Chung, T.-S.; Helmer, B.J.; De Wit, J.S. Enhanced double-skinned FO membranes with inner dense layer for wastewater treatment and macromolecule recycle using Sucrose as draw solute. *J. Membr. Sci.* **2012**, *396*, 92–100. [CrossRef]
12. Yen, S.K.; Mehnas Haja, N.F.; Su, M.; Wang, K.Y.; Chung, T.-S. Study of draw solutes using 2-methylimidazole-based compounds in forward osmosis. *J. Membr. Sci.* **2010**, *364*, 242–252. [CrossRef]
13. Miller, D.G.; Lee, C.M.; Rard, J.A. Ternary Isothermal Diffusion Coefficients of NaCl-MgCl$_2$-H$_2$O at 25 °C. 7. Seawater Composition. *J. Solut. Chem.* **2007**, *36*, 1559–1567. [CrossRef]
14. Li, G.; Li, X.-M.; He, T.; Jiang, B.; Gao, C. Cellulose triacetate forward osmosis membranes: Preparation and characterization. *Desalin. Water Treat.* **2013**, *51*, 2656–2665. [CrossRef]
15. Lim, S.; Akther, N.; Phuntsho, S.; Shon, H.K. Defect-free outer-selective hollow fiber thin-film composite membranes for forward osmosis applications. *J. Membr. Sci.* **2019**, *586*, 281–291. [CrossRef]
16. Xie, M.; Nghiem, L.D.; Price, W.E.; Elimelech, M. Comparison of the removal of hydrophobic trace organic contaminants by forward osmosis and reverse osmosis. *Water Res.* **2012**, *46*, 2683–2692. [CrossRef]
17. Gao, Y.; Li, W.; Lay, W.C.L.; Coster, H.G.L.; Fane, A.G.; Tang, C.Y. Characterization of forward osmosis membranes by electrochemical impedance spectroscopy. *Desalination* **2013**, *312*, 45–51. [CrossRef]
18. Lotfi, F.; Phuntsho, S.; Majeed, T.; Kim, K.; Han, D.S.; Abdel-Wahab, A.; Shon, H.K. Thin film composite hollow fibre forward osmosis membrane module for the desalination of brackish groundwater for fertigation. *Desalination* **2015**, *364*, 108–118. [CrossRef]
19. Liu, Y.; Mi, B. Combined fouling of forward osmosis membranes: Synergistic foulant interaction and direct observation of fouling layer formation. *J. Membr. Sci.* **2012**, *407–408*, 136–144. [CrossRef]
20. Yoon, H.; Baek, Y.; Yu, J.; Yoon, J. Biofouling occurrence process and its control in the forward osmosis. *Desalination* **2013**, *325*, 30–36. [CrossRef]
21. Li, Z.Y.; Yangali-Quintanilla, V.; Valladares-Linares, R.; Li, Q.; Zhan, T.; Amy, G. Flux patterns and membrane fouling propensity during desalination of seawater by forward osmosis. *Water Res.* **2012**, *46*, 195–204. [CrossRef] [PubMed]
22. Jeong, S.; Kim, S.-J.; Hee Kim, L.; Seop Shin, M.; Vigneswaran, S.; Vinh Nguyen, T.; Kim, I.S. Foulant analysis of a reverse osmosis membrane used pretreated seawater. *J. Membr. Sci.* **2013**, *428*, 434–444. [CrossRef]
23. Kang, G.-D.; Gao, C.-J.; Chen, W.-D.; Jie, X.-M.; Cao, Y.-M.; Yuan, Q. Study on hypochlorite degradation of aromatic polyamide reverse osmosis membrane. *J. Membr. Sci.* **2007**, *300*, 165–171. [CrossRef]
24. Holloway, R.W.; Maltos, R.; Vanneste, J.; Cath, T.Y. Mixed draw solutions for improved forward osmosis performance. *J. Membr. Sci.* **2015**, *491*, 121–131. [CrossRef]

25. Majeed, T.; Phuntsho, S.; Sahebi, S.; Kim, J.E.; Yoon, J.K.; Kim, K.; Shon, H.K. Influence of the process parameters on hollow fiber-forward osmosis membrane performances. *Desal. Wat. Treat.* **2015**, *54*, 817–828. [CrossRef]
26. Mi, B.; Elimelech, M. Chemical and physical aspects of organic fouling of forward osmosis membranes. *J. Membr. Sci.* **2008**, *320*, 292–302. [CrossRef]
27. She, Q.; Jin, X.; Li, Q.; Tang, C.Y. Relating reverse and forward solute diffusion to membrane fouling in osmotically driven membrane processes. *Water Res.* **2012**, *46*, 2478–2486. [CrossRef]
28. Zhao, S.; Zou, L. Relating solution physicochemical properties to internal concentration polarization in forward osmosis. *J. Membr. Sci.* **2011**, *379*, 459–467. [CrossRef]
29. Lee, S.; Boo, C.; Elimelech, M.; Hong, S. Comparison of fouling behavior in forward osmosis (FO) and reverse osmosis (RO). *J. Membr. Sci.* **2010**, *365*, 34–39. [CrossRef]
30. Liyanaarachchi, S.; Jegatheesan, V.; Obagbemi, I.; Muthukumaran, S.; Shu, L. Effect of feed temperature and membrane orientation on pre-treatment sludge volume reduction through forward osmosis. *Desalin. Water Treat.* **2015**, *54*, 838–844. [CrossRef]
31. Valladares Linares, R.; Yangali-Quintanilla, V.; Li, Z.; Amy, G. Rejection of micropollutants by clean and fouled forward osmosis membrane. *Water Res.* **2011**, *45*, 6737–6744. [CrossRef] [PubMed]
32. Cath, T.Y.; Gormly, S.; Beaudry, E.G.; Flynn, M.T.; Adams, V.D.; Childress, A.E. Membrane contactor processes for wastewater reclamation in space: Part I. Direct osmotic concentration as pretreatment for reverse osmosis. *J. Membr. Sci.* **2005**, *257*, 85–98. [CrossRef]
33. Xie, M.; Nghiem, L.D.; Price, W.E.; Elimelech, M. A forward osmosis-membrane distillation hybrid process for direct sewer mining: System performance and limitations. *Environ. Sci. Technol.* **2013**, *47*, 13486–13493. [CrossRef] [PubMed]
34. Cornelissen, E.R.; Harmsen, D.; de Korte, K.F.; Ruiken, C.J.; Qin, J.-J.; Oo, H.; Wessels, L.P. Membrane fouling and process performance of forward osmosis membranes on activated sludge. *J. Membr. Sci.* **2008**, *319*, 158–168. [CrossRef]
35. Achilli, A.; Cath, T.Y.; Marchand, E.A.; Childress, A.E. The forward osmosis membrane bioreactor: A low fouling alternative to MBR processes. *Desalination* **2009**, *239*, 10–21. [CrossRef]
36. Holloway, R.W.; Childress, A.E.; Dennett, K.E.; Cath, T.Y. Forward osmosis for concentration of anaerobic digester centrate. *Water Res.* **2007**, *41*, 4005–4014. [CrossRef]
37. Xie, M.; Nghiem, L.D.; Price, W.E.; Elimelech, M. Toward resource recovery from wastewater: Extraction of phosphorus from digested sludge using a hybrid forward osmosis-membrane distillation process. *Environ. Sci. Technol. Lett.* **2014**, *1*, 191–195. [CrossRef]
38. Sun, W.; Shi, J.; Chen, C.; Li, N.; Xu, Z.; Li, J.; Lv, H.; Qian, X.; Zhao, L. A review on organic–inorganic hybrid nanocomposite membranes: A versatile tool to overcome the barriers of forward osmosis. *RSC Adv.* **2018**, *8*, 10040. [CrossRef]
39. Shaffer, D.L.; Werber, J.R.; Jaramillo, H.; Lin, S.; Elimelech, M. Forward Osmosis: Where are we now? *Desalination* **2015**, *356*, 271–284. [CrossRef]
40. Shan, M.; Kang, H.; Xu, Z.; Li, N.; Jing, M.; Hu, Y.; Teng, K.; Qian, X.; Shi, J.; Liu, L. Decreased cross-linking in interfacial polymerization and heteromorphic support between nanoparticles: Towards high-water and low-solute flux of hybrid forward osmosis membrane. *J. Colloid Interface Sci.* **2019**, *548*, 170–183. [CrossRef]
41. Kang, H.; Wang, W.; Shi, J.; Xu, Z.; Lv, H.; Qian, X.; Liu, L.; Jing, M.; Li, F.; Niu, J. Interlamination restrictive effect of carbon nanotubes for graphene oxide forward osmosis membrane via layer by layer assembly. *Appl. Surface Sci.* **2019**, *465*, 1103–1106. [CrossRef]

© 2020 by the authors. Licensee MDPI, Basel, Switzerland. This article is an open access article distributed under the terms and conditions of the Creative Commons Attribution (CC BY) license (http://creativecommons.org/licenses/by/4.0/).

Article

Tailoring the Effects of Titanium Dioxide (TiO$_2$) and Polyvinyl Alcohol (PVA) in the Separation and Antifouling Performance of Thin-Film Composite Polyvinylidene Fluoride (PVDF) Membrane

Shruti Sakarkar [1,*], Shobha Muthukumaran [2] and Veeriah Jegatheesan [1]

1 School of Engineering, RMIT University, Melbourne, VIC 3000, Australia; jega.jegatheesan@rmit.edu.au
2 College of Engineering and Science, Victoria University, Melbourne, VIC 8001, Australia; shobha.muthukumaran@vu.edu.au
* Correspondence: shruti.sakarkar@gmail.com; Tel.: +61-404-041-643

Citation: Sakarkar, S.; Muthukumaran, S.; Jegatheesan, V. Tailoring the Effects of Titanium Dioxide (TiO$_2$) and Polyvinyl Alcohol (PVA) in the Separation and Antifouling Performance of Thin-Film Composite Polyvinylidene Fluoride (PVDF) Membrane. *Membranes* **2021**, *11*, 241. https://doi.org/10.3390/membranes11040241

Academic Editors: Tae-Hyun Bae and Pei Sean Goh

Received: 15 February 2021
Accepted: 24 March 2021
Published: 28 March 2021

Publisher's Note: MDPI stays neutral with regard to jurisdictional claims in published maps and institutional affiliations.

Copyright: © 2021 by the authors. Licensee MDPI, Basel, Switzerland. This article is an open access article distributed under the terms and conditions of the Creative Commons Attribution (CC BY) license (https://creativecommons.org/licenses/by/4.0/).

Abstract: In this study, thin-film composite (TFC) polyvinylidene fluoride (PVDF) membranes were synthesized by coating with titanium dioxide (TiO$_2$)/polyvinyl alcohol (PVA) solution by a dip coating method and cross-linked with glutaraldehyde. Glutaraldehyde (GA) acted as a cross-linking agent to improve the thermal and chemical stability of the thin film coating. The incorporation of TiO$_2$ in the film enhanced the hydrophilicity of the membrane and the rejection of dyes during filtration. The layer of TiO$_2$ nanoparticles on the PVDF membranes have mitigated the fouling effects compared to the plain PVDF membrane. The photocatalytic performance was studied at different TiO$_2$ loading for the photodegradation of dyes (reactive blue (RB) and methyl orange (MO)). The results indicated that the thin film coating of TiO$_2$/PVA enhanced photocatalytic performance and showed good reusability under UV irradiation. This study showed that nearly 78% MO and 47% RB were removed using the TFC membrane. This work provides a new vision in the fabrication of TFC polymeric membranes as an efficient wastewater treatment tool.

Keywords: dip-coating; dyes; membrane; nanocomposite; polyvinylidene fluoride; titanium dioxide

1. Introduction

Rapid urbanization and industrialization are leading to water scarcity and deterioration of the quality of freshwater. Effluents discharged from industries contain persistent organic pollutants (POP), which are synthetic chemicals with pronounced persistence against chemical or biological degradation and bioaccumulation that have substantial impacts on human health and the environment, even at very minimal concentrations over a prolonged period [1].

Synthetic textile dyes are emerging POPs from the textile industries [2]. The textile industry is the highest water consuming industry and generates a large volume of effluent during the dyeing and finishing processes, where most of the dyes were washed out with water [3]. Effluents generated by the textile mills are notorious for their complexity, comprising synthetic dyes, cleansing agents, salts, surfactants, dispersants, inhibitory compounds, oil, toxic chemicals, and many other compounds [4,5]. The chemicals present in synthetic dyes can cause fatal and mutagenic effects on living beings and aquatic life [6]. Therefore, the main challenge is to reduce and remediate contaminated textile effluents economically and in a sustainable manner. Traditional technologies used to treat textile effluents have limitations and are still not adequate for complete degradation and removal of dye residues, as most of the compounds are highly resilient for these processes to be effective [7]. Among different wastewater treatments, the membrane filtration process has attracted significant attention for textile effluents' treatment because of its high separation efficiency, simplistic operation, no sludge production, and easy to scale-up. Despite the advantages mentioned above, fouling is the major factor that restricts the application of polymeric membranes for textile effluent treatment [8]. The incorporation of nanoparticles into membranes is a novel approach to overcome the disadvantages of fouling in polymeric membranes.

Thin-film composite (TFC) membrane with nanoparticle incorporated on the superficial layer improves the physico-chemical, mechanical, and thermal properties of the membrane. The advantage of the superficial layer is that it acts as a selective barrier during the separation process, which enhances the separation performance of the membrane. Various types of nanoparticles have been incorporated into the top layer of TFC membranes, such as carbon nanotubes [9], graphene oxide [10], silica [11], zeolites [12], and titanium dioxide [13–16]. Additional selective transport channels can be formed in TFC membranes due to that porous nanofiller or the interfaces between nanofiller and polymer matrices, facilitating the separation property of the resultant membrane.

Previous studies have shown that TFC membranes have great potential in separating and reducing pollutants from wastewater [17,18]. The incorporation of nanoparticles in the thin film enhances the water permeability, antibacterial, and antifouling properties of the membrane [19].

Titanium dioxide (TiO_2) is probably the most widely used nanomaterial in membrane modification [20,21] due to their ability to detoxify harmful organic pollutants through their high photocatalytic activity, larger surface area, flexibility in the surface function, and their mechanical stability towards the UV irradiation [22]. Various methods used to immobilize the TiO_2 nanoparticles for the surface modification of a membrane are dip or spin coating, blending, hot pressing, and physical and chemical cross-linking [23,24].

Cross-linking of TiO_2 nanoparticles using polymers, such as polyvinyl alcohol (PVA) onto a membrane have been found to have significant application in wastewater treatment [25–27]. The idea behind PVA application is to introduce a hydrophilic group into the polymeric membrane to reduce the membrane's fouling [28]. The oleophobic behavior of PVA decreases the fouling rate and enhances the membrane's thermal and mechanical properties. The incorporation of TiO_2 improves photodegradation performance due to the election-hole separation when irradiated by UV light with an energy equal to or greater than the bandgap energy of the TiO_2 nanoparticles [29]. In our previous work, polyvinylidene fluoride (PVDF) membranes were synthesized by optimizing the PVA content at a fixed TiO_2 concentration. The optimization was carried out with respect to the water flux and removal of dyes achieved by the membranes [30]. The study showed that 3 wt.% of PVA was the optimum loading for the TFC of the membrane. Li et al. [31] observed synthesized PVDF and PVA hollow fiber membrane modified with TiO_2 nanoparticles enhanced the dye rejection ability of the membrane compared to the PVA thin film membrane. Liu et al. [32] demonstrated that the thin film membrane composed of PVA and TiO_2 nanoparticles showed improved photoactivity under visible light [30,33]. All these findings demonstrate that the combination of TiO_2 and PVA coating of TFC membranes improves the physio-chemical properties of the membrane compared to the conventional polymeric membranes.

This study aims to fabricate a TFC membrane with TiO_2 nanoparticles incorporated into the PVA layer to improve the antifouling and dye rejection performance. The concentration of TiO_2 is optimized in this study. The surface chemistry, surface morphology and hydrophilicity of the TiO_2/PVA incorporated membranes were also characterized at different TiO_2 loading.

Moreover, the separation capability of TFC membrane was studied using the cross-flow filtration method and ex-situ method for photodegradation of methyl orange (MO) and reactive blue (RB) dyes. The effect of different parameters, such as pH, TiO_2 loading, and zeta potential, were studied to understand the behavior of the TFC membranes. Furthermore, the kinetic study of photodegradation of dyes by TFC membranes conducted at different TiO_2 loadings is considered one of the novelties of this paper.

In addition, the stability of the thin film coating is an important parameter that decides the separation performance of the membrane. The stability of the PVA/TiO$_2$ coating during multiple filtration runs have not been reported so far. Therefore, in this study, we have focused on the stability of the TFC membranes under UV irradiation during multiple runs.

2. Materials and Methods

2.1. Materials

All the chemicals used for this research were analytical grade and used without any further purification. Pellet of PVDF (MW = 534,000 g/mol) and PVA (99% hydrolyzed) were obtained from Sigma–Aldrich (NSW, Australia). Cross-linking agent glutaraldehyde (25% in H$_2$O) was procured from Sigma–Aldrich (NSW, Australia). MO and RB were used for the separation, whereas bovine serum albumin (BSA) was used to prepare the feed solution to study membrane fouling, and all of them were purchased from Sigma–Aldrich (NSW, Australia). The type, molecular weight, and chemical structures of the dyes used as well as the wavelength at which maximum absorption of those dyes occurred are given in Table 1. TiO$_2$ (Aeroxide® P25, (Brunauer–Emmett–Teller, BET) surface area = 35–65 m^2/g, particle size ≈ 21 nm) were acquired from Sigma–Aldrich (NSW, Australia). N-N dimethylacetamide (DMAc), Sulphuric acid (H$_2$SO$_4$), sodium hydroxide (NaOH), hydrochloric acid (HCl), and isopropanol (IPA) were acquired from Merck chemicals (Melbourne, Australia).

Table 1. Physio-chemical properties of the dyes used in this study.

Dye	Abbreviation	Chemical Formula	Chemical structure	Molecular Weight (g/mol)	Type	Wavelength at Which the Maximum Absorbance Occurred, λ_{max} (nm)
Methyl orange	MO	C$_{14}$H$_{14}$N$_3$NaO$_3$S		626.50	Azo dye	464
Reactive blue	RB	C$_{22}$H$_{16}$N$_2$Na$_2$O$_{11}$S$_3$		327.33	Anthraquinone dye	590

2.2. Methods

2.2.1. Fabrication of PVDF Flat Sheet Membranes

A flat sheet PVDF membrane was fabricated using the non-solvent induced phase separation (NIPS) technique. Dried PVDF (16 wt.%) pellets were mixed mechanically with DMAc (84 wt.%) solvent overnight to obtain a homogeneous solution. Then the solution was sonicated at 60 °C for 30 min for complete degassing. The casting solution is then spread uniformly on a glass plate using a thin film applicator (Elecometer 4340, Manchester, England) at a thickness of 200 µm; then the film was immersed in a non-solvent bath containing a mixture of water (70 vol%) and isopropanol (30 vol%) for the precipitation at room temperature for at least 10 min. Further details of the membrane preparation are described elsewhere [34].

2.2.2. Surface Modification of PVDF Membranes Using TiO$_2$/PVA Solution

TFC PVDF membranes were prepared by chemical cross-linking reaction of PVA and TiO$_2$ with GA. PVA was dissolved in water at 100 °C until a transparent solution was obtained, followed by adding of TiO$_2$ nanoparticles (Table 2). Viscosities of the solutions were measured by a stress-controlled rheometer (Reologica Merlin II Rheosys). PVDF membrane was washed with a mixture of ethanol and water and spread evenly on a glass plate for complete drying. PVA/TiO$_2$ sol was then poured on a clean and dried PVDF membrane and spread evenly all over it. Excess solution was removed by hanging membrane at room temperature (as shown in Appendix A Figure A1). Figure 1 demonstrates the flow diagram for the preparation and synthesis of the TFC PVDF membranes. Lastly, TFC membranes were dried at room temperature before cross-linking.

Table 2. Composition of TFC membranes and the viscosity of coating solutions (TiO$_2$/PVA).

Membrane (Abbreviation)	Modified Solution Composition (TiO$_2$: PVA) (wt.%/wt.%)	Viscosity (Pa. s)
Membrane 1 (PT0)	Plain PVDF	
Membrane 2 (PT1)	1:3	0.7 ± 0.1
Membrane 3 (PT2)	1.5:3	1.3 ± 0.2
Membrane 4 (PT3)	2:3	1.5 ± 0.1
Membrane 5 (PT4)	3:3	2.4 ± 0.2
Membrane 6 (PT5)	5:3	3.7 ± 0.3

(in all TFC membranes, PVDF = 16 wt.% and DMAc = 84 wt.%)

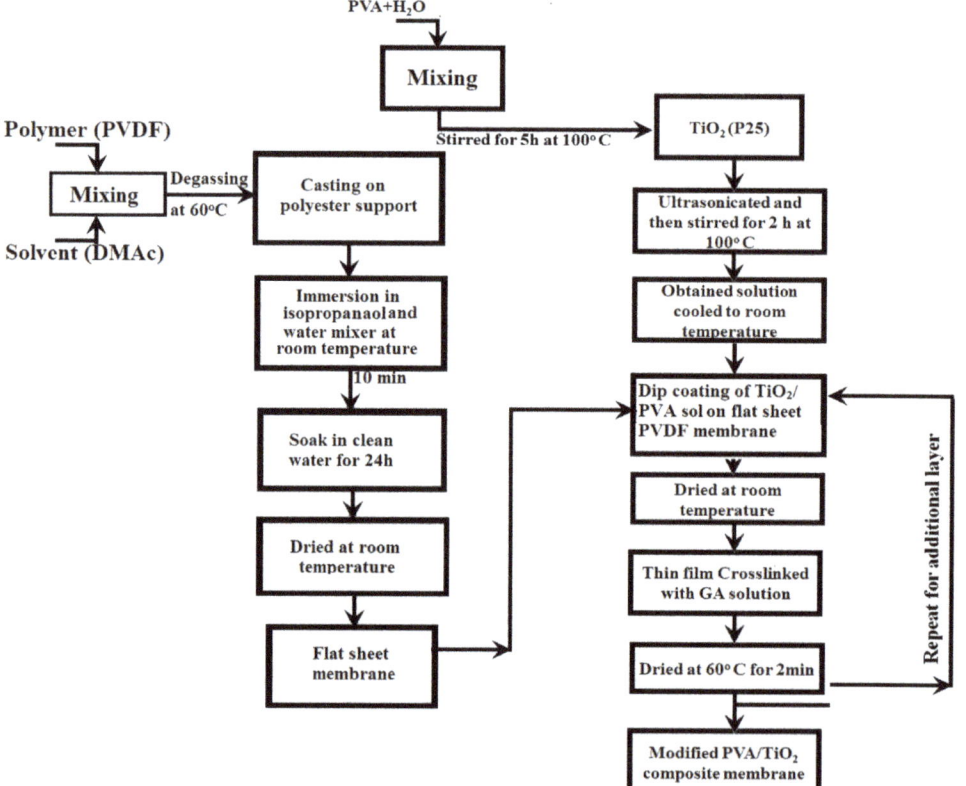

Figure 1. Preparation and synthesis flow diagram for the synthesis of TFC TiO$_2$/PVA/ PVDF membrane.

TFC TiO$_2$/PVA/PVDF membrane was soaked in a cross-linking solution consisting of 2 mL GA and 0.5 mL sulfuric acid in water for 5 min at room temperature. The excess solution was removed, and the membrane was dried at 60 °C for 2 min for better cross-linking. Here sulfuric acid acted as a catalyst for the process of cross-linking. The selection of GA and sulfuric acid was based on our previous work recommendations [30]. Each membrane was coated twice to obtained better efficacy for TiO$_2$ nanoparticles.

2.3. Membrane Characterization

2.3.1. Surface Morphology and Chemical Composition of TiO$_2$/PVA Composite PVDF Membrane

The plain and modified TFC PVDF membrane's surface morphology were examined using a field-emission in-lens scanning electron microscope (FEISEM) (FEI Nova Nano SEM, Hillsboro, OR, USA). Washed and dried membranes were cut into small pieces and sputter-coated with gold particles for electrical conductance. The images were scanned under high voltage at different magnification to understand the presence and dispersion of TiO$_2$ nanoparticles over the membrane's surface. Energy dispersive X-ray (EDX) (Oxford X-Max20 EDX detector) was used to study the membrane's elemental composition. In this study, EDX was carried out in conjunction with the SEM by using AZtecAM software. Each elemental composition was estimated using the average of three measurements of each sample.

Surface roughness for all the membranes were measured using an atomic force microscope (AFM) (Asylum MFP-3D Infinity). Small pieces (approximately 1.5 cm × 1.5 cm) of the prepared membranes were glued on a flat and hard substrate such as glass. The membrane was scanned and imaged in a scan size of 2 µm × 2 µm. Roughness parameters were obtained and reported in terms of root mean square roughness (RMS) and mean surface roughness (Sa); values are calculated by using Gwyddion software (Version 1.2). Recorded results are the average of three measurements of a sample.

The functional affinity of plain and TFC PVDF membranes were analyzed using Fourier-transform infrared spectroscopy (FTIR) (Thermo FTIR spectrometer, Dreieich, Germany).

The crystalline phases of the membranes with and without thin film coating were compared using high-resolution X-ray diffraction (XRD) (BRUKER-AXS, Karlsruhe, Germany) with the integrated software diffract EVA (Version 09.2017).

The membrane's hydrophilicity at different TiO$_2$ loading in thin-film coating was calculated by the sessile drop method using the OCA20 instrument (Scientex, Melbourne, Australia) at room temperature. Images were captured when droplet fell on the dry membrane surface, and then contact angles with water were measured by supplied software. The reported contact angle is the average of five consecutive measurements of the same membrane at different locations.

2.3.2. Surface Charge of the Membrane

The membrane surface Zeta potential was determined by a streaming potential analyzer (Malvern surface zeta potential). The membrane's surface charge was measured by adding a small piece (1 mm × 1 mm) of a membrane in milli-Q water and at a required pH. The reported values are the average of three runs of three different samples of each membrane at pH ranging from 3 to 11. The chemicals such as NaOH (base) and HCl (acid) were used to adjust the milli-Q water's pH.

2.3.3. Filtration Performance and Fouling Analysis

The filtration performance of the plain and TFC PVDF membranes' were analyzed through cross-flow filtration set up (Figure 2). A cross-flow filtration system carried out all the filtration experiments for the membrane with different TiO$_2$ loading. The membrane with an active filtration area of 0.00746 m^2 was placed in the cross-flow filtration unit, and the flow rate was kept constant at 18 L/h at an operating pressure of 3 bar. Milli-Q water was used for the calculation of pure water flux by using Equation (1).

$$J = \frac{V}{A \times t} \quad (1)$$

where V is the total volume of permeate collected within time t, and A is the active membrane surface area used for filtration. The two different dyes (RB and MO) were used to prepared model dye solutions at a concentration of 50 mg/L each. The feed solution was circulated throughout the setup for at least 20 min before the reject was collected for the analysis. All filtration experiments were carried out at room temperature.

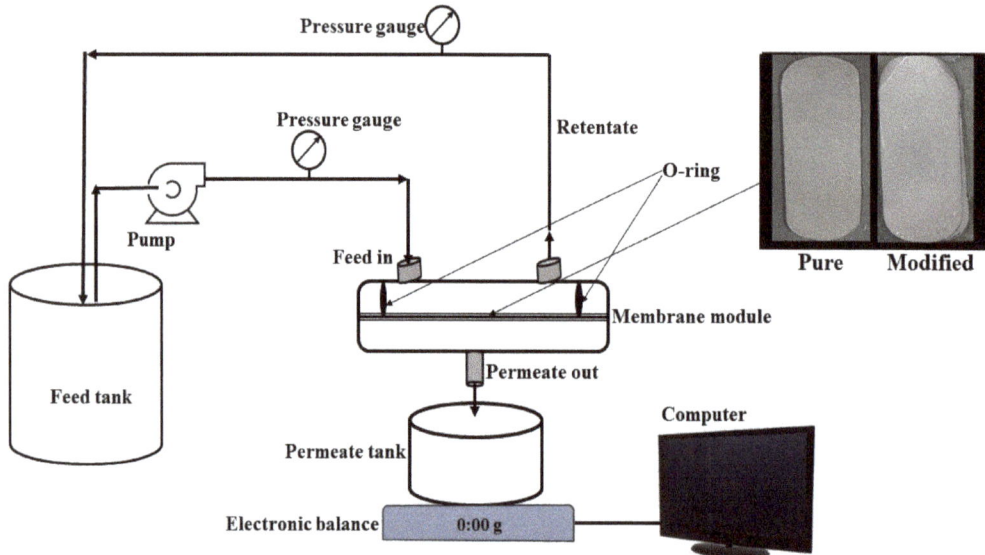

Figure 2. Schematic representation of the cross-flow filtration unit.

Samples (feed and permeate) were collected at regular intervals, and change in the concentration of dyes were measured by UV-Vis spectrometer (Shimadzu UV 2700, Kyoto, Japan). The wavelength at which the maximum absorbance of a dye obtained was found for each dye solution to produce calibration curves and to find the concentration of dyes in the samples collected. The rejection percentage, R (%), for each experiment was calculated using Equation (2) given below:

$$\text{R (\%)} = \left(1 - \frac{C_p}{C_f}\right) \times 100 \tag{2}$$

where C_p is the permeate concentration, and C_f is the concentration of feed.

2.3.4. Analysis of Membrane Fouling

Fouling effects on plain and TFC PVDF membranes were analyzed as described below:
First, the cross-flow filtration of pure water was carried out for 30 min using the selected membrane. This was followed by the filtration of 1 g/L of BSA solution for another 60 min. The feed BSA solution was recirculated during this period. Then, the cross-flow filtration of pure water was resumed and continued for another 50 min. Permeates of pure water and BSA solution were collected continuously to compute flux changes during this period. The membrane resistance was calculated using the pure water flux obtained before and after the BSA solution's filtration. The irreversible fouling (IF) factor of the membrane was calculated using Equation (3).

$$\text{IF} = \frac{J_0 - J_1}{J_0} \times 100 \tag{3}$$

where J_0 and J_1 are the pure water flux values before and after the filtration of BSA solution.

2.3.5. Measurement of Photocatalytic Activity

Photocatalytic effectiveness of the TFC PVDF membranes were evaluated by photodegradation of the synthetic dyes (MO and RB) in an aqueous solution. All photocatalytic experiments were studied by the ex-situ method. The prepared nanocomposite membrane (5 cm × 5 cm) was immersed in 130 mL of 50 mg/L of an aqueous solution of a dye and was kept in the dark for 30 min to stabiles the adsorption of dye onto the membrane surface. The solution was then irradiated under two Ultraviolet-C (UV-C) lamps (Philips TUV 15W/G15 T8) with a light intensity of 2.1 mW/cm^2 each. Samples from the solution were collected at regular intervals to observe the photodegradation of dyes. Change in concentration was examined using UV-spectrometer. Photodegradation (D) was calculated in percentages of dye degradation according to Equation (4):

$$\text{Photodegradation, } D(\%) = \frac{C_i - C_f}{C_i} \times 100 \qquad (4)$$

where C_i and C_f are the initial and final concentrations of dye before and after UV irradiation.

3. Results and Discussions
3.1. The Surface Morphology of TFC Membranes at Different TiO$_2$ Loading

The surface morphology of the plain and TFC PVDF membranes were investigated using SEM analysis. As illustrated in Figure 3(a,a'), the plain PVDF membrane's surface has some obvious macro-pores, whereas the modified membranes showed a layer of TiO$_2$/PVA over the substrate. The distribution of TiO$_2$ nanoparticles was non-uniform at smaller concentrations (i.e., at 1, 1.5, and 2 wt.% TiO$_2$); in contrast, the further increase in TiO$_2$ loading has led to a uniform distribution over the membrane surface and part of that agglomerated on the surface which can be clearly seen in the SEM images (Figure 3). When the TiO$_2$ loading was increased beyond 2 wt.%, the nanoparticles might have incorporated into the membrane's pores, which will reduce the membrane's porosity. Porosities of the membranes have been studied in one of our previous works for the membrane containing PVA/TiO$_2$ thin film using gravimetric analysis [30]. The aggregation or stacking of nanoparticles may induce a noticeable change in the photocatalytic performance and the stability of the composite membranes. The aggregation of nanoparticles increases the thickness of the TiO$_2$/PVA thin film, which decreases the permeate flux.

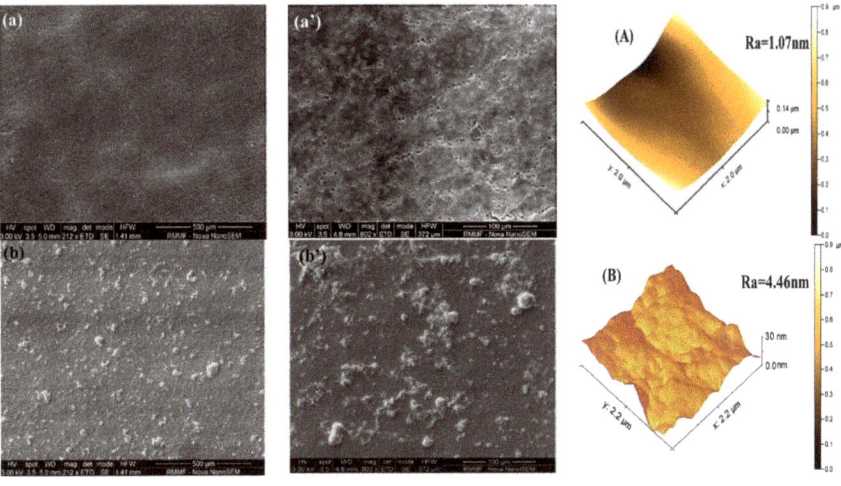

Figure 3. Cont.

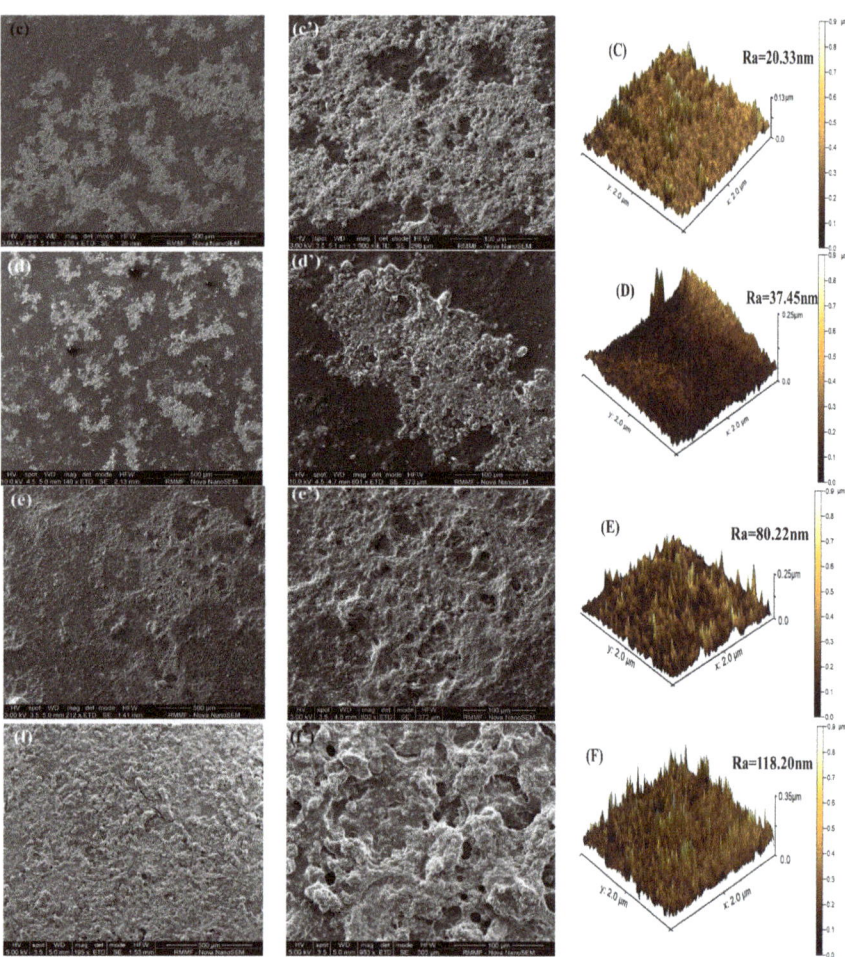

Figure 3. SEM images of the top surface of the TFC membranes at different magnification at different TiO_2 loading (**a,a′**) = 0 wt.% TiO_2; (**b,b′**) = 1 wt.% TiO_2; (**c,c′**) = 1.5 wt.% TiO_2; (**d,d′**) = 2 wt.% TiO_2; (**e,e′**) = 3 wt.% TiO_2; (**f,f′**) = 5 wt.% TiO_2 and AFM images of the corresponding membranes (**A**) = 0 wt.% TiO_2; (**B**) = 1 wt.% TiO_2; (**C**) = 1.5 wt.% TiO_2; (**D**) = 2 wt.% TiO_2; (**E**) = 3 wt.% TiO_2 and (**F**) = 5 wt.% TiO_2.

Table 2 indicates that an increase in TiO_2 loading increases the solution's viscosity, which leads to the formation of a thick film over the substrate. Figure 4A,B illustrates the cross-sectional images of plain and TFC membranes with 1 wt.% TiO_2. The TiO_2 coated membrane has two distinct layers, one with a thin film of TiO_2/PVA, having a thickness of 5 μm, which did not exist in the plain PVDF membrane. When the solution viscosity increases, the layer's thickness also increased due to the agglomeration of nanoparticles in the solution. The presence of the TiO_2 nanoparticles on the membrane surface was confirmed by the Energy-dispersive X-ray spectroscopy (EDX) analysis (Figure 4C,D). The main peak of 0.67 and 0.277 keV represent the fluorine and carbon peak, which are abundant because they signify the PVDF membrane. A peak detected at 4.5 keV belongs to Ti, and 0.52 keV represents oxygen indicating the presence of the TiO_2 nanoparticles in the composite membrane.

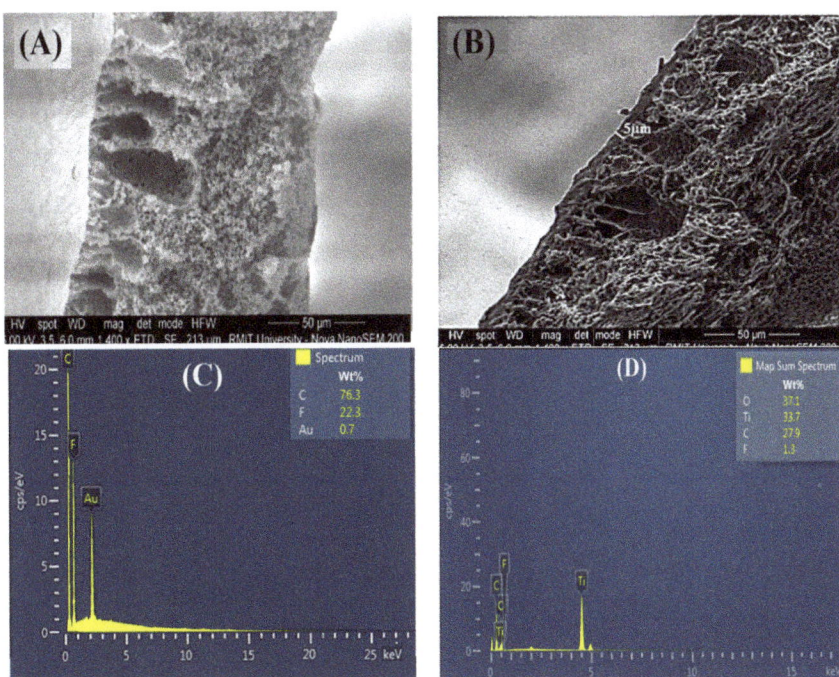

Figure 4. The SEM images of cross-sections of plain and modified PVDF membranes: (**A**) plain PVDF membrane (0 wt.% TiO$_2$) and (**B**) TiO$_2$/PVA modified PVDF membrane (1 wt.% TiO$_2$); and the corresponding EDX images of the top surface of the membrane (**C**) plain PVDF membrane (0 wt.% TiO$_2$) and (**D**) TiO$_2$/PVA modified PVDF membrane (1 wt.% TiO$_2$).

3.2. Surface Roughness Analysis of TiO$_2$ Composite PVDF Membrane

The surface topography is one of the powerful techniques for mapping the surface morphology, roughness, and adhesive properties of the nanoparticles on the PVDF membrane with and without modification. Figure 3A–F represents the AFM images of the plain and TiO$_2$/PVA coated membranes. The average value of root means square (RMS) and the average roughness (Ra) of different membranes are presented in Table 3. It can be observed that the TiO$_2$ loading has increased the surface roughness of the modified membrane. AFM images show that the membrane surface possesses peak and valley like structures (the brightest parts in each AFM images represent the peak, whereas the darkest parts represent the valleys). The increasing surface roughness with increasing TiO$_2$ loading confirms the adhesion of TiO$_2$ nanoparticles on the TFC membrane's surface. The increase in TiO$_2$ loading has also increased the peak and valley, which could help to enhance photocatalysis. When a dye solution passed through the membrane, these valleys can adsorb the dye molecules, which could be successfully degraded by the hydroxyl radicals produced by TiO$_2$ under UV irradiation [35].

Table 3. Surface roughness parameters at different TiO$_2$ loading on the PVDF membranes obtained from AFM images (Figure 3).

Membranes	Root Mean Square Roughness (RMS) (nm)	Mean Roughness (Sa) (nm)
PT0	1.07	0.9
PT1	4.46	3.5
PT2	20.33	13.4
PT3	37.45	24.40
PT4	80.40	45.43
PT5	118.20	102.5

3.3. The Contact Angle of the Membrane

The wettability of the samples plays a very decisive role in determining the water permeability of a membrane. Generally, it is considered as the higher the degree of hydrophilicity, the greater the water permeability [36]. The contact angles of the plain and modified PVDF membranes are illustrated in Figure 5A. The plain PVDF membrane contact angle is 87.42° and decreased with the increase in TiO_2 loading. Membrane with 5 wt.% TiO_2 shows the lowest contact angle 48°. The increase in TiO_2 concentration also implies the increase in PVA on the membrane surface, which creates the hydroxyl groups on the surface [31]. The increase in the hydroxyl group enhances the hydrophilicity of the membrane. Mänttäri et al. [37] found that a membrane with a lower contact angle exhibit a significant change in zeta potential with an increase in pH.

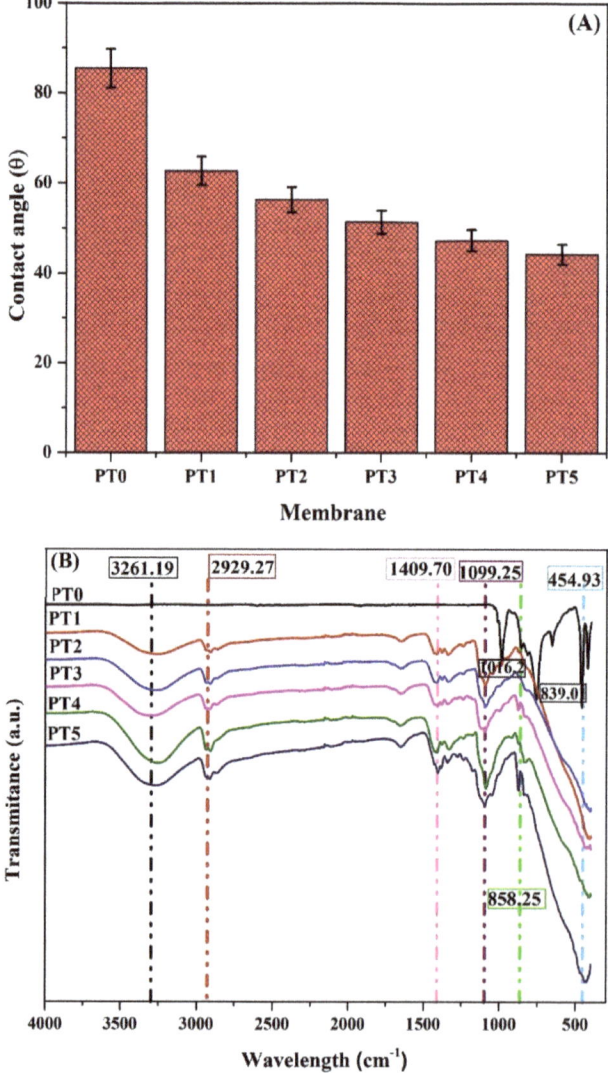

Figure 5. Cont.

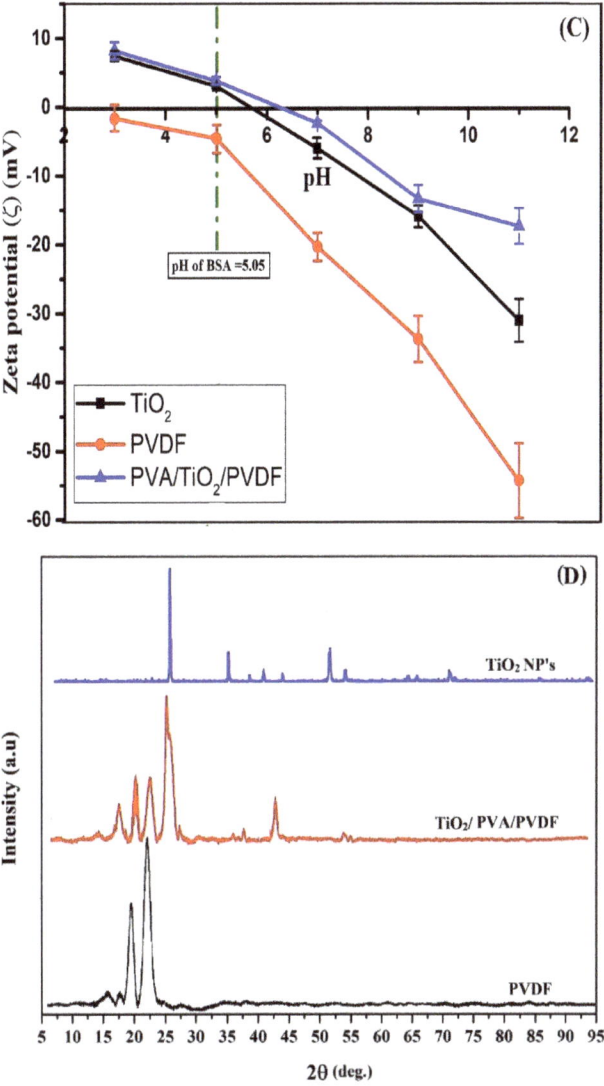

Figure 5. Effect of TiO$_2$ loading (**A**) Contact angle of the membranes; (**B**) FTIR spectra of plain and modified PVDF membranes; (**C**) Zeta potential values at different pH for TiO$_2$, plain PVDF and 1 wt.% TiO$_2$/PVA modified PVDF membrane; (**D**) XRD pattern of TiO$_2$ nanoparticles, plain and modified (for 1 wt.% TiO$_2$) PVDF membranes.

3.4. FTIR Analysis

Fourier-transform infrared (FTIR) spectroscopy was used to study the change in the functional group on the membrane surface after the modification by TiO$_2$/PVA thin film coating (with increasing TiO$_2$ loading). The measurements were taken from 4000 to 400 cm^{-1}, and the functional groups that participated in the structure were determined from the peaks of the graphs of transmission against the wavelength. In Figure 5B, the absorption band at 1140–1280 cm^{-1} and 1411–1419 cm^{-1} is characteristic stretching of CF$_2$ and CH$_2$ group, respectively, found in the PVDF membrane [38]. The PVDF membrane also showed peaks at 873.88, 832.88, and 476.86 cm^{-1} which have been associated with the

C-C-C bond. A distinct peak at 2929.27, 1409.70, and 1099.25 cm^{-1} resulted in the stretching of -CH$_2$, C=O, and C-O, respectively, which are the characteristics peak of PVA [39]. A broad peak in the TiO$_2$/PVA composite membrane at 2900–3500 cm^{-1} represents the Ti-OH group, which shows the presence of -OH hydrophilic group on the surface and indicates the presence of TiO$_2$ nanoparticles [40]. The increase in TiO$_2$ loading enhanced the intensity of the peak at 3258.95 cm^{-1}. The increase in peak intensity of -OH band represents the increase in TiO$_2$ nanoparticles on the surface of all the membranes (PT1 to PT5).

3.5. Surface Charge of the Membrane

The surface charge has a significant effect on the removal efficiency and membrane fouling. The surface charge can be characterized in terms of zeta potential, as illustrated in Figure 5C. Zeta potential (ζ) changed with pH of the solution due to protonation and deprotonation of the membrane's functional group. Its values, either positive or negative, have a substantial effect on stabilizing the particles in the suspension. This is attributed to the electrostatic repulsion between particles with the same electric charge that causes the particles' segregation [41]. The PVDF membrane is negatively charged over the entire pH range due to the C-F group's electronegative charge [42], whereas TiO$_2$ particles show positive zeta potential in acidic conditions and start to fall off with the increase in pH. The zeta potential of TiO$_2$/PVA coated PVDF membrane is positive in acidic conditions and becomes negative in alkaline conditions. The acquired positive charge in acidic conditions indicates the presence of TiO$_2$ on the PVDF membrane. The decrease in the zeta potential value with the increase in pH is due to the surface adsorption of OH$^-$ and Cl$^-$ anions from the solution.

3.6. Crystalline Structure of the Synthesized Membranes

Crystallinity is one of the important parameters which affects the chemical and mechanical properties of a polymer. The XRD diffraction patterns of TiO$_2$ nanoparticles, plain PVDF membrane, and TiO$_2$/PVA coated PVDF membrane containing 1 wt.% of TiO$_2$ nanoparticles are shown in Figure 5D. TiO$_2$ is a mixture of two different forms (75% anatase and 25% rutile); therefore, the TiO$_2$ pattern has three dominating peaks at 2θ = 25.25°, 37.12°, and 48.05° [43,44]. The diffraction peaks at 2θ = 20.8° and 18.56° are the characteristics of α-phase PVDF polymer, which were also observed in the TiO$_2$/PVA modified. It consists of characteristic peaks at 19.5° and 38.6°, which are related to the semi-crystalline nature of PVA [18,30,45]. The diffraction peaks at 25.2 and 45.89 indicate the anatase structure of TiO$_2$ [46]. It indicates that nano-TiO$_2$ particles did not change the crystal structure of TiO$_2$ on the surface after modification.

3.7. Performance and Photocatalytic Activity of the Membrane

3.7.1. Flux and Removal of Dyes by Composite Membranes

The variation in pure water flux of membranes coated with TiO$_2$ (0, 1, 1.5, 2, 3, 5 wt.%) nanoparticles was evaluated at 3 bar and presented in Figure 6A. The incorporation of TiO$_2$ on the membrane surface results in blocking the surface pores and causing the reduction in the permeate flux [47]. SEM images also show that an increase in the loading of TiO$_2$ nanoparticles improves the dispersions with the formation of a thin film of TiO$_2$/PVA over the PVDF membrane's surface, which results in the decrease in pore size. A gradual reduction in pure water flux was observed as the TiO$_2$ loading was increased in the membranes from PT1 to PT5. The pure water flux for all the membranes is comparatively higher than the permeate flux obtained for dye solutions due to the accumulation of dye molecules on the membrane surface, causing membrane fouling and therefore reducing the membrane flux. Comparing to RB dye, the permeate flux obtained for MO dye solution was slightly higher. Lower molecular weight dye (MO) can easily pass through the membrane compared to the higher molecular weight dye (RB), which can accumulate on the membrane surface and yielding lower permeate flux.

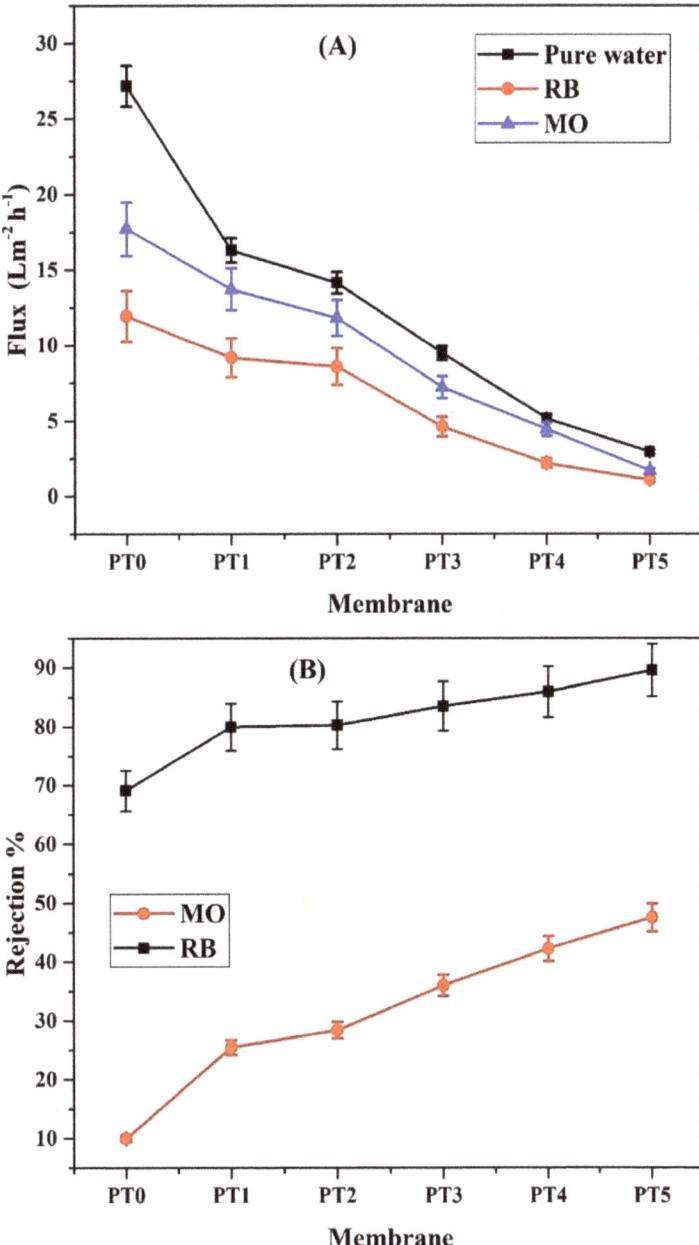

Figure 6. Cont.

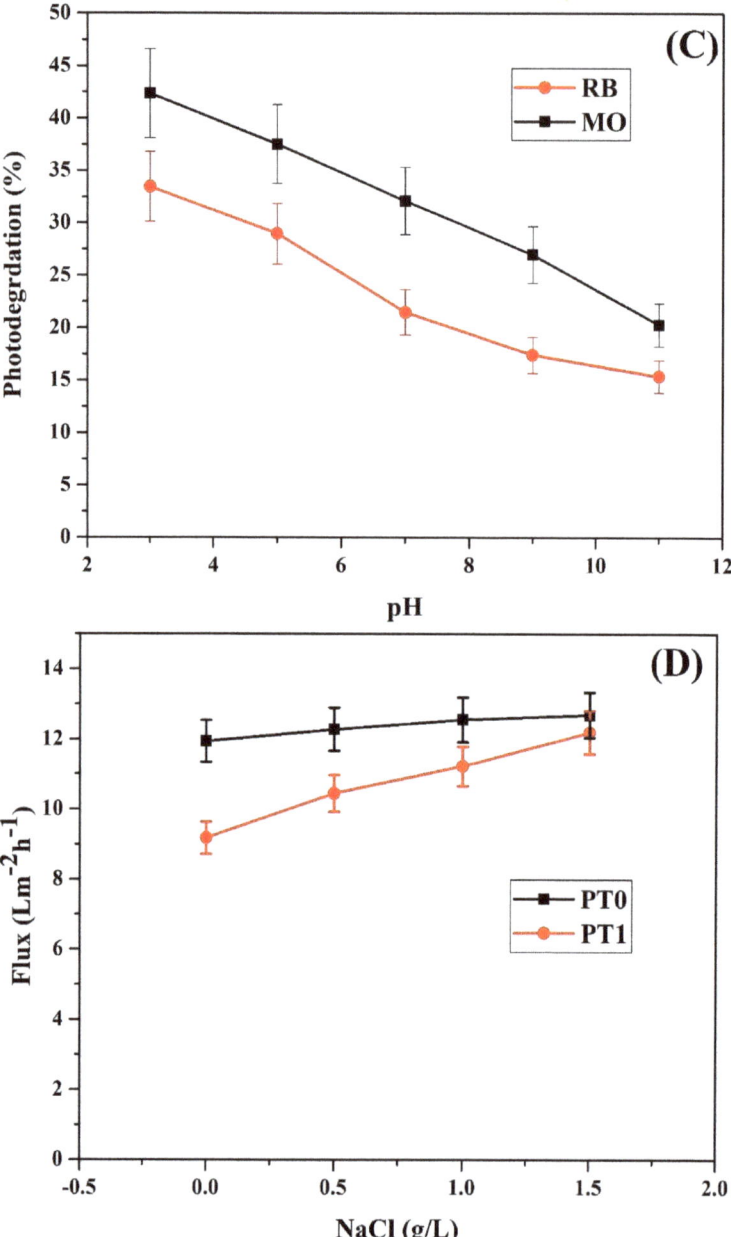

Figure 6. *Cont.*

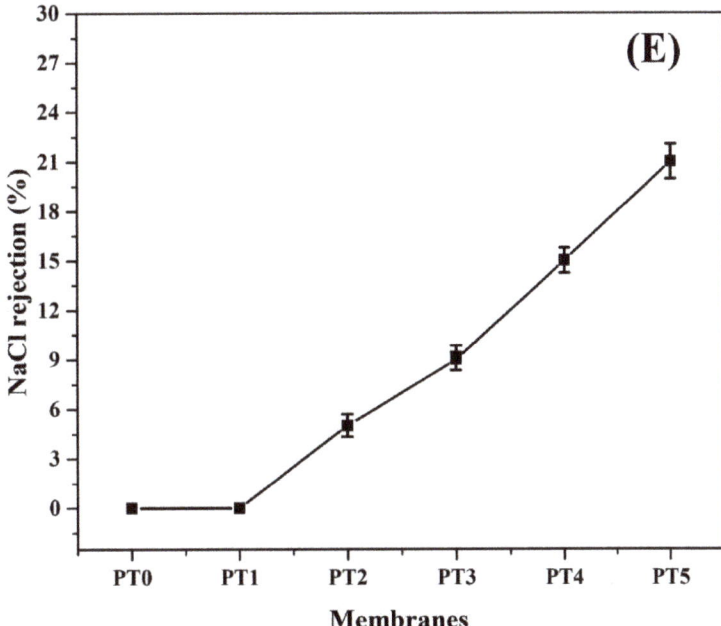

Figure 6. (**A**) Average flux for modified and unmodified PVDF membranes at different TiO_2 loading (0 to 5 wt.%) at an operating pressure of 3 bar for pure water and the dye solutions (RB and MO) (Dye concentration = 50 mg/L); (**B**) Rejection of dyes by TiO_2/PVA modified and unmodified PVDF membrane during filtration (Dye concentration = 50 mg/L and experimental duration = 60 min);(**C**) Effect of solution pH on photodegradation of dyes (MO and RB) at 50 mg/L of dye concentrations under UV irradiation (Time =180 min); (**D**) Effect of NaCl concentration on RB (50 mg/L) treatment (P = 0.3 MPa); (**E**) NaCl rejection at different TiO_2 loading.

Rejection of different dyes with membranes coated with different TiO_2 concentrations (0 to 5 wt.%) was studied at a temperature of 27 ± 3 °C at an operating pressure of 3 bar. The concentrations were 50 mg/L for both dyes, and the filtration was continued for 60 min. Figure 6B indicates that the modified membrane with TiO_2/PVA improved dye rejection compared to the plain PVDF membrane. The rejection values better for RB dye due to its higher molecular weight, whereas MO dye showed low rejection values. It implies that the dye with higher molecular weight can be adsorbed easily onto the membrane surface [48]. Almost complete removal of dye was obtained in the case of PT5 membrane but with the lowest permeate flux of 0.95 $Lm^{-2}h^{-1}$.

3.7.2. Effect of Solution pH on TFC Membrane

pH is considered an important parameter in determining the removal performance and efficiency of the photocatalytic process. The influence of solution pH in the removal of RB and MO was studied by changing the solution pH from 3 to 11, and the results are shown in Figure 6C. Tests were conducted with an initial concentration of 50 mg/L of RB and MO, at a minimum TiO_2 loading (1 g/L) under UV irradiation. According to Figure 6C, RB and MO degradation is higher in acidic conditions and decreased in alkaline conditions. In the feed solution, the dye molecule state is changed due to pH, which influenced the rate of photodegradation [31]. In Figure 5C, at pH less than 5.9, the surfaces of TiO_2 nanoparticles are positively charged, whereas changed to negative charge at pH higher than 5.9. Thus, pH lower than that corresponding to the point of zero charge supports the adsorption of dye (RB and MO) molecules on the surface of the membrane, which improves the degradation of dyes under acidic and neutral condition.

Furthermore, Nakabayashi et al. [49] explained that at lower pH value, TiO_2 nanoparticles could produce higher concentrations of hydroxyl ions, which subsequently cause an increased photocatalytic degradation. Additionally, the increase in pH value increases the coulomb repulsion between the negative surface charge of the TFC membrane and the OH radicals in the photocatalytic oxidation leading to a decrease in photodegradation [50].

3.7.3. Effect of Salt on TFC Membrane

Textile mills commonly use salts for the dying process and most of which are discharged in the effluent and can have adverse effects on the dye removal process. Therefore, the effect of NaCl salt with RB dye solution (50 mg/L) was studied with and without PVA/TiO_2 coated PVDF membrane (Figure 6D). An increase in the salt concentration increased the permeate flux rate for the TiO_2/PVA coated membrane compared to that for the plain PVDF membrane. This is due to the Donnan effect between the surface groups (hydroxyl group) of the TiO_2/PVA coated membrane and the ions with the same charge weakened with the increase in salt concentration and increased flux [10]. The increasing salt concentration could cause thinning of PVA/TiO_2 layer on the membrane surface due to charge interactions [31].

The effect of TiO_2 loading on the removal of salt is represented in Figure 6E. The NaCl rejection increased from 0% for PT0 to 21% for PT5. The deposition of TiO_2/PVA film on the PVDF membrane's surface might increase the film thickness, which lowers the flux but enhances the salt rejection. The obtained results indicate that the increasing TiO_2 loading in the TFC membrane improves permeability and ion rejection.

3.7.4. Fouling Study of PVA/TiO_2 TFC PVDF Membrane

The fouling behavior of the TiO_2/PVA PVDF membrane was carried out using the BSA solution (1 g/L) for 60 min at a pressure of 3 bar. The data from Figure 7A indicates that the initial flux of plain PVDF membrane (PT0) was higher, but the start to decline during the BSA solution's filtration. Whereas the TiO_2 coated membrane shows similar flux before and during the BSA rejection. The results reveal that the antifouling properties of TiO_2/PVA coated membrane has improved considerably compared to that of the plain PVDF membrane. Moreover, the increase in TiO_2 loading enhanced the fouling mitigation properties of the membrane. Polymer surfaces are more reactive to protein molecules than TiO_2 molecules; therefore, incorporating TiO_2 in PVDF membrane prevents the adsorption of protein on the membrane surface [51]. The presence of TiO_2 nanoparticles helps to wash away the protein molecules from the composite membranes compared to the plain PVDF membrane. Wang et al. [52] have observed that the accumulation of TiO_2 nanoparticles boosted the electron donor mono-polarity of the TiO_2/PVA composite membranes and enhanced the repulsive interaction energy between foulants and membrane surfaces to improve the antifouling ability.

Further, the zeta potential of the plain and modified PVDF membrane containing 1 wt.% TiO_2 at pH 5 were −4.56 and 3.78 mV, respectively. The pH of the BSA solution was also around 5, and thus both the modified PVDF membrane and BSA might have the same surface charge (positive), leading to the resistance by the membrane to adsorb BSA.

The pure water flux recovery for the PT1 membrane was 76.2% after filtering BSA through it, whereas for the plain PVDF membrane, the pure water flux recovery was about 38.9%. The results also indicate that the irreversible fouling factors for the modified (PT1) and unmodified (PT0) PVDF membranes were 25 and 61%, respectively.

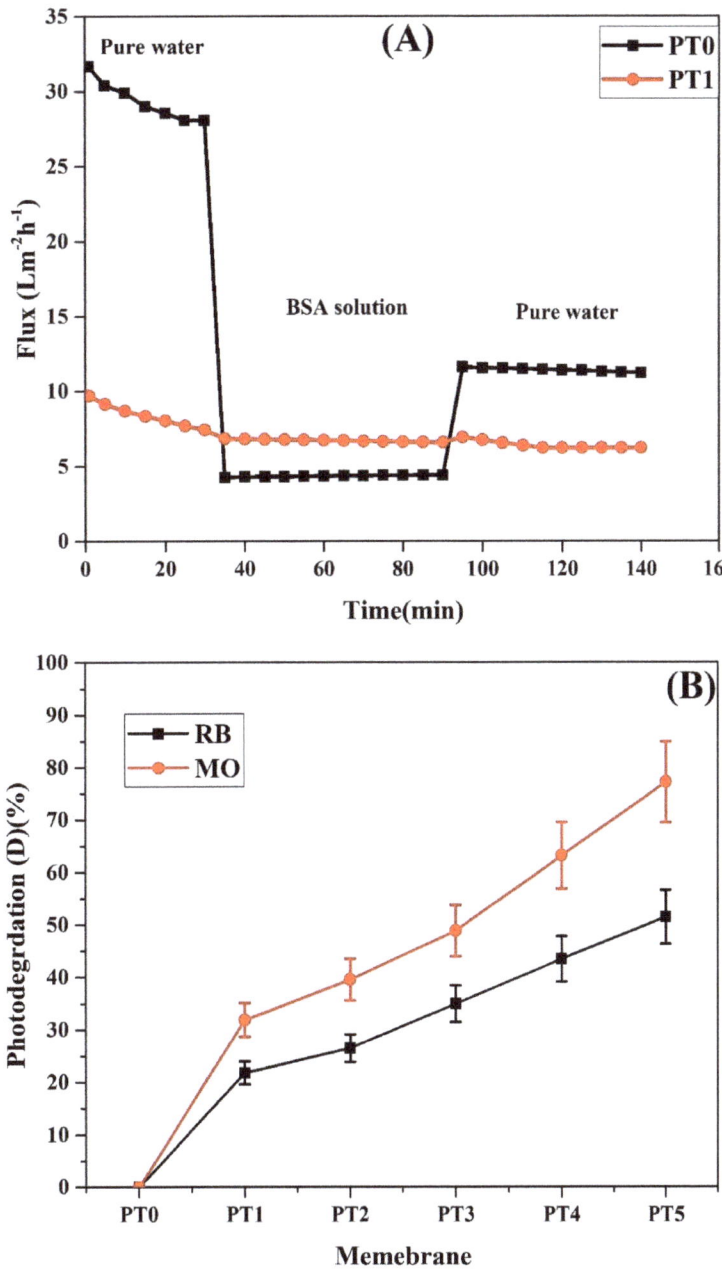

Figure 7. (**A**) Bovine serum albumin (BSA) removal using plain and modified membranes (PT0 and PT1) (BSA concentration = 1 g/L, operating pressure = 3 bar); (**B**) Photocatalytic degradation of dyes (reactive blue and methyl orange) at a dye concentration of 50 mg/L under UV irradiation for 150 min.

3.7.5. Photocatalytic Performance and Degradation Kinetics at Different TiO$_2$ Concentration

Figure 7B illustrates the effect of TiO$_2$ loading on the reactive blue and methyl orange dye's photodegradation at a concentration of 50 mg/L. The photocatalytic performance of the membranes was evaluated by measuring the change in the concentration of dyes for a duration of 180 min (for the initial 30 min, the membrane was soaked in the dye solution and kept in the dark, and for the rest of the time, the membrane was soaked in the dye solution and exposed to UV irradiation). The results indicated that the adsorption of the dyes was less than 5% in the dark.

The temporal change in the concentration of RB and MO in the presence of different membranes with and without TiO$_2$ are illustrated in Figure 8A,C. Results indicate that the membranes with TiO$_2$ nanoparticles show significant photocatalytic activity compared to that of the plain PVDF membrane. The plain PVDF membrane does not show any removal under UV irradiation, whereas the dye degradation rate increases for the TiO$_2$/PVA coated membrane. The increase in TiO$_2$ loading (0 to 5 wt.%) enhanced the photo degradability of the dyes due to the generation of the hydroxyl group. The higher the concentration of TiO$_2$, the higher the hydroxyl radicals produced, which enhanced the degradation rate. The maximum degradation of 77.2% for MO and 51.5% for RB were achieved under UV irradiation for the PT5 membrane. This was due to the generation of active radical species such as positive holes (h$^+$), hydroxyl radicals (OH$^\bullet$), and superoxide radicals (O$_2$$^\bullet$) on the surface of the modified membranes due to the incorporation of TiO$_2$ nanoparticles. Those radicals attack the dye molecules, causing their degradation into intermediates through various reaction followed by the complete mineralization into less toxic CO$_2$ and H$_2$O.

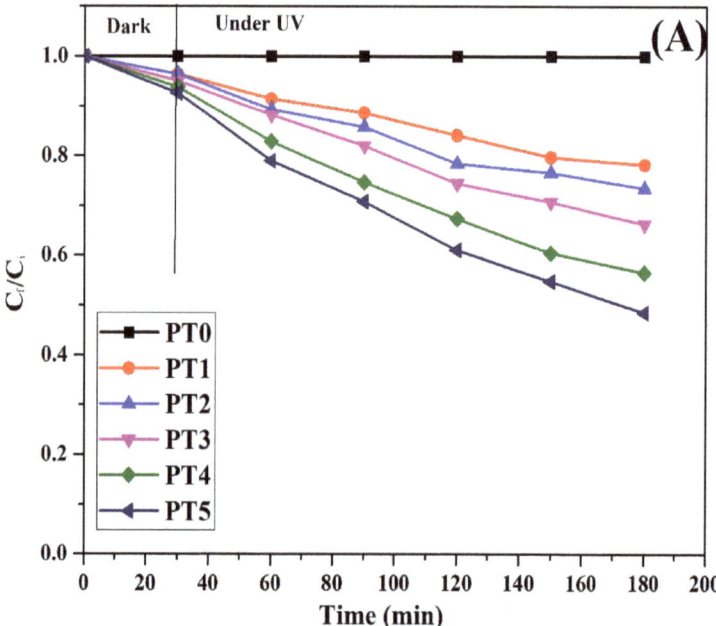

Figure 8. *Cont.*

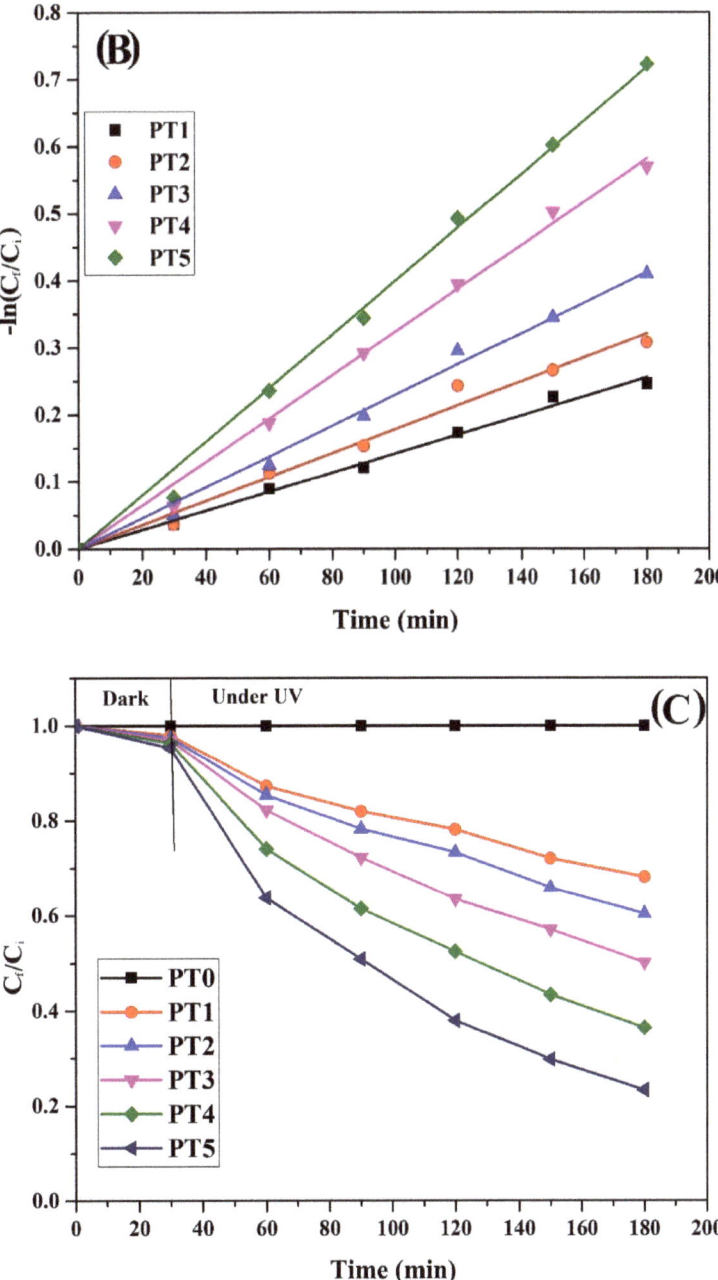

Figure 8. *Cont.*

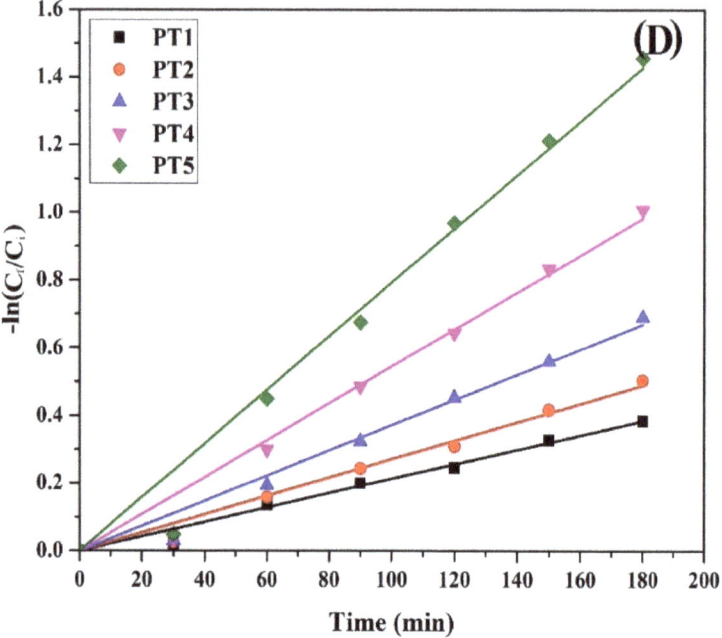

Figure 8. Temporal variation of normalized dye concentration in the permeate obtained from the plain and the modified PVDF membranes: (**A**) RB (**C**) MO; Natural log plots of normalized dye concentration in the permeate obtained from the plain and the modified PVDF membranes against time to compute rate constants: (**B**) RB (**D**) MO.

To quantify the photocatalytic effects on the degradation of dyes at different TiO_2 loading, the Langmuir-Hinshelwood (L-H) kinetic model was used [53].

$$-\frac{dC}{dt} = \frac{kKC}{1+KC} \quad (5)$$

After integrating the above equation, we can obtain the following relationship:

$$\ln\left(\frac{C_i}{C_f}\right) + K(C_i - C_f) = kKt \quad (6)$$

where C_i, C_f, are the initial and final concentrations of the dye at time 0 and t, respectively; k and K are the reaction rate constant and adsorption equilibrium constant, respectively. C is the concentration at a given time, t. The order of Equation (5) will be zero when C_f is relatively high, and KC is $\gg 1$. When the concentration of dye in the solution decreases, then KC will become $\ll 1$. Thus, the KC in the denominator of Equation (5) will be negligible, and the reaction will be the first order. Overall, dyes' degradation rate was increased with the increase in the concentration of TiO_2 nanoparticles over the surface of the membrane. Thus, assuming the first-order decay of dyes, the following equation can be used:

$$-\frac{dC}{dt} = kKC = k'C \quad (7)$$

Integrating the Equation (7) will yield Equation (8):

$$-\ln\frac{C_f}{C_i} = k't \quad (8)$$

The linear regressions obtained from the plots $-\ln(C_f/C_i)$ v/s time is shown in Figure 8B,D, and the values of the rate constants are listed in Table 4. The half-reaction time, $t_{1/2}$ required for the concentration to drop one-half of its initial concentration can be given by Equation (9):

$$t_{1/2} = \frac{\ln(2)}{k'} \tag{9}$$

Table 4. The apparent first-order rate constant (k') and half-life time of MO and RB degradation by plain and modified PVDF membranes.

Membrane	MO			RB		
	Rate Constant k' (min^{-1})	R^2	$t_{1/2}$ (min)	Rate Constant k' (min^{-1})	R^2	$t_{1/2}$ (min)
PT1	0.0021	0.993	330	0.0014	0.997	495
PT2	0.0027	0.992	256.66	0.0017	0.994	407.64
PT3	0.0037	0.991	187.29	0.0022	0.997	315
PT4	0.0054	0.991	128.33	0.0032	0.998	216.56
PT5	0.0079	0.991	87.72	0.0039	0.998	177.69

The results indicate that the increase in TiO_2 concentration in the film enhanced the reaction rate and decreased the half-reaction time. The shortest half reaction time corresponds to the high rate of degradation.

3.7.6. The Mechanisms of Photodegradation of Dyes by TFC Membrane

The detailed mechanism of the degradation and the antifouling process of the TFC PVDF membrane is explained below (Figure 9). The mechanisms of photodegradation are photoexcitation, migration, oxidation-reduction reaction, and charge separation [45]. When the TiO_2 coated surface is exposed to UV irradiation (hυ) with the energy equal or higher than the bandgap energy, electrons are excited from valence band (VB) to conduction band (CB), generating holes (h_{vb}^+) at VB and electrons (e_{cb}^-) at CB. Holes and electrons react with species adsorbed on the surface of the TiO_2 catalyst. Valence band holes react with water (H_2O/OH^-) to generate hydroxyl radicals (HO•), while electrons react with adsorbed molecular oxygen (O_2) and reducing it to superoxide radical anion, which, in turn, reacts with protons to form peroxide radicals [54,55] The relevant reaction that occurs on the surface of the TFC PVDF membrane for the photodegradation of the RB and MO can be described as:

$$TiO_2 + h\upsilon \ (\lambda \leq 365 \text{ nm}) \rightarrow h_{(vb)}^+ + e_{(cb)}^- \tag{10}$$

The electron from CB is readily surrounded by the O_2 to generate superoxide radicals ($O_2^{•-}$) on the surface of TiO_2

$$e_{(cb)}^- + O_2 \rightarrow (O_2^{•-}) \tag{11}$$

The generated superoxide radicals ($O_2^{•-}$) could react with the water molecules (H_2O) to produce hydroxyl radicals (OH•) and hydroperoxyl radicals ($HO_2^•$) which are the strong oxidizing agents to degrade the organic molecules.

Whereas the holes produced on VB surrounded by the hydroxyl group (H_2O) to generate hydroxyl radicals (OH•) on the surface of TiO_2.

$$h_{(vb)}^+ + OH^- \rightarrow OH^• \tag{12}$$

In an aqueous medium, VB holes can react with the surface adsorbed water molecules to form hydroxyl species.

$$Ti - H_2O + h_{(vb)}^+ \rightarrow Ti - OH^• + H^+ \tag{13}$$

Therefore, when the dye (RB and MO) molecules react with the hydroxyl radicals, they oxidized into CO_2 and H_2O as follows;

$$C_{14}H_{14}N_3NaO_3S \text{ (MO) or } C_{22}H_{16}N_2Na_2O_{11}S_3 \text{ (RB)} + OH^\bullet + O_2 \rightarrow \text{Product } (CO_2) \text{ and } (H_2O) + \text{degradation product} \quad (14)$$

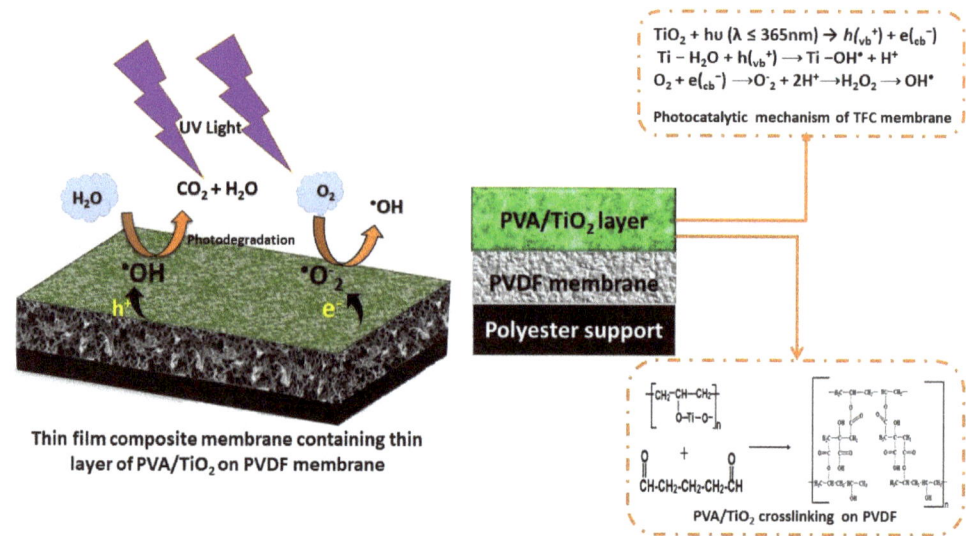

Figure 9. Photocatalytic mechanism of TFC membrane under UV irradiation.

Further, our main focus is to modify the PVDF membranes and understand morphological behavior. The absorption of the dyes was studied under dark condition and with UV irradiation. The change in concentration was calculated by observing the change in absorption at a maximum wavelength (MO = 464 and RB = 590). As per Deng et al. [56], many organic dyes such as erythrosine, eosin Y, rhodamine B, and rose bengal have been utilized to sensitize catalysts for H_2 production. Despite MO and RB are organic dyes, they are not utilized for catalyst sensitization. Further studies are warranted in this area.

3.8. Stability of TiO_2 Nanoparticles Coated onto the Membrane Surface during Photocatalysis

The stability of the TiO_2 nanoparticles over the surface of the membrane has an essential role in the membrane's performance. The long-term stability of the modified membrane has been tested by degrading RB repeatedly three times. In order to understand the stability of TiO_2 nanoparticles on the TiO_2/PVA composite PVDF membrane, the membrane was reused for the photodegradation of the RB dye. TiO_2/PVA composite membranes were successively used three times for the dye degradation; after each use, the membranes were washed with milli-Q water and dried at room temperature. A fresh batch of dye solution (50 mg/L) was prepared for the next cycle, and the experiment was performed under the same operating conditions. Figure 10 shows that the photodegradation rate decreased in the second and third runs, compared to that in the first run. The membrane with higher TiO_2 (PT5) showed a change in photodegradation of RB dye (~24 %) after the third cycle. Therefore, we conclude that the TiO_2/PVA thin film coating is suitable for the degradation of dyes under UV irradiation. The loss in photocatalytic efficiency is mainly due to the loss of TiO_2 nanoparticles during the washing of membranes [57].

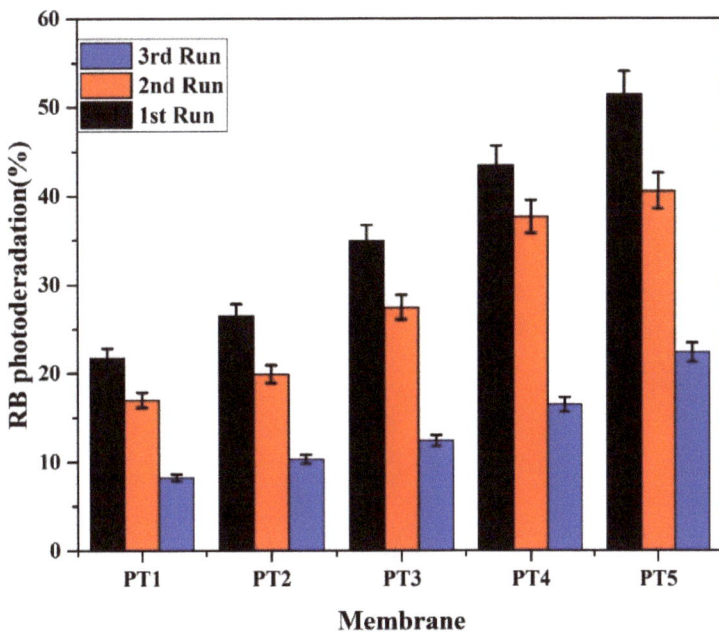

Figure 10. Degradation performance (stability) of TiO$_2$ nanoparticles while degrading reactive blue dye under UV irradiation during repeated runs (Dye concentration = 50 mg/L, 150 min. of UV irradiation per run).

4. Conclusions

In this research, the photocatalytic activity of the TFC PVDF membrane was studied to treat the aqueous solution of reactive blue and methyl orange dyes. Membranes synthesized by dip coating at different TiO$_2$ loading were characterized using spectroscopic techniques. Based on the SEM, FTIR, and XRD measurements, the presence of TiO$_2$ nanoparticles on the surface of the membrane was confirmed. PVA helps to obtain better efficacy of TiO$_2$ nanoparticles over the PVDF membrane's surface. Glutaraldehyde is a cross-linker for the PVA polymer chains to enhance the thin film mechanical and chemical stability. The TFC membrane showed better removal of dyes, enhanced antifouling ability, and hydrophilicity compared to the plain PVDF membrane. The enhancements of surface negative charge and hydrophilicity made the TFC membrane more resistant to the deposition of dyes. The increasing concentration of TiO$_2$ nanoparticles in thin-film coating improved the photocatalytic activity and the rate of degradation of dyes. It was found that the degradation of RB and MO fitted the first-order kinetics.

As the TiO$_2$ loading increases in the thin film, nanoparticles start to aggregate, reducing the porosity and the permeate flux of the membrane. The stability of nanoparticles remained for three experimental cycles with a slight decrease in photodegradation capacity. The attractive results provide valuable insights into the design and fabrication of high-performance polymeric TFC membrane and demonstrate a promising approach for the efficient removal of dyes in an aqueous solution.

Author Contributions: Conceptualization, S.S. and V.J.; methodology, S.S., S.M., and V.J.; software, S.S.; validation, S.S. and V.J.; formal analysis, S.S.; investigation, S.S.; resources, S.S.; data curation, S.S.; writing—original draft preparation, S.S.; writing—review and editing, S.S., S.M., and V.J.; supervision, V.J. and S.M.; project administration S.S. and V.J. All authors have read and agreed to the published version of the manuscript.

Funding: This research received no external funding.

Institutional Review Board Statement: Not applicable

Informed Consent Statement: Not applicable

Data Availability Statement: Data is contained within the article.

Conflicts of Interest: The authors declare no conflict of interest. The funders had no role in the design of the study; in the collection, analyses, or interpretation of data; in the writing of the manuscript, or in the decision to publish the results.

Appendix A

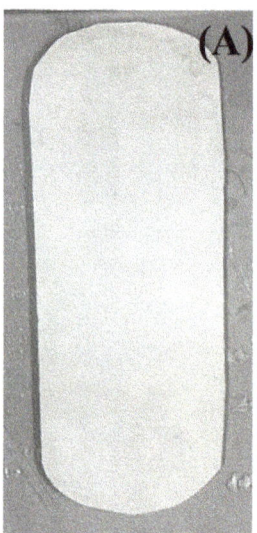 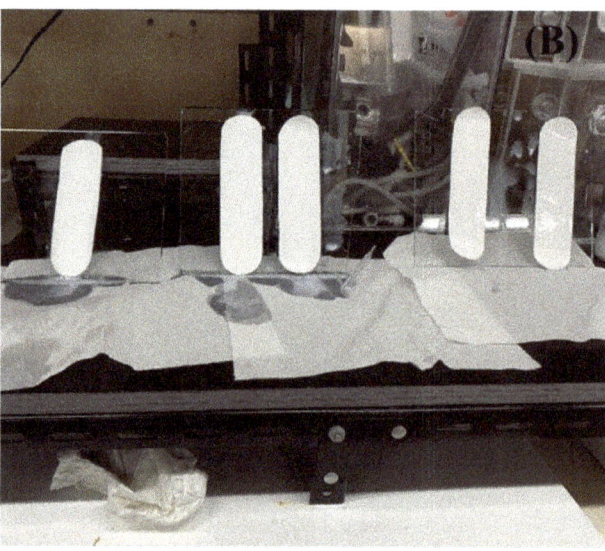

Figure A1. Actual images of the membrane (**A**) neat PVDF membrane (**B**) TiO_2 /PVA coated PVDF membrane with different TiO_2 loading (hang them for the removal of excess solution.

References

1. Mohanapriya, S.; Mumjitha, M.; Purnasai, K.; Raj, V. Fabrication and characterization of poly (vinyl alcohol)-TiO_2 nanocomposite films for orthopedic applications. *J. Mech. Behav. Biomed. Mater.* **2016**, *63*, 141–156. [CrossRef]
2. Do Vale-Júnior, E.; da Silva, D.R.; Fajardo, A.S.; Martínez-Huitle, C.A. Treatment of an azo dye effluent by peroxi-coagulation and its comparison to traditional electrochemical advanced processes. *Chemosphere* **2018**, *204*, 548–555. [CrossRef]
3. Homaeigohar, S. The nanosized dye adsorbents for water treatment. *Nanomaterials* **2020**, *10*, 295. [CrossRef]
4. Holkar, C.R.; Jadhav, A.J.; Pinjari, D.V.; Mahamuni, N.M.; Pandit, A.B. A critical review on textile wastewater treatments: Possible approaches. *J. Environ. Manag.* **2016**, *182*, 351–366. [CrossRef]
5. Hayat, H.; Mahmood, Q.; Pervez, A.; Bhatti, Z.A.; Baig, S.A. Comparative decolorization of dyes in textile wastewater using biological and chemical treatment. *Sep. Purif. Technol.* **2015**, *154*, 149–153. [CrossRef]
6. Khan, S.; Malik, A. Environmental and health effects of textile industry wastewater. In *Environmental Deterioration and Human Health*; Springer: Dordrecht, The Netherlands, 2014; pp. 55–71. [CrossRef]
7. Tehrani-Bagha, A.R.; Mahmoodi, N.M.; Menger, F.M. Degradation of a persistent organic dye from colored textile wastewater by ozonation. *Desalination* **2010**, *260*, 34–38. [CrossRef]
8. Li, J.; Yuan, S.; Zhu, J.; Van der Bruggen, B. High-flux, antibacterial composite membranes via polydopamine-assisted PEI-TiO_2/Ag modification for dye removal. *Chem. Eng. J.* **2019**, *373*, 275–284. [CrossRef]
9. Kim, E.S.; Hwang, G.; El-Din, M.G.; Liu, Y. Development of nanosilver and multi-walled carbon nanotubes thin-film nanocomposite membrane for enhanced water treatment. *J. Membr. Sci.* **2012**, *394*, 37–48. [CrossRef]
10. Lai, G.S.; Lau, W.J.; Goh, P.S.; Ismail, A.F.; Yusof, N.; Tan, Y.H. Graphene oxide incorporated thin film nanocomposite nanofiltration membrane for enhanced salt removal performance. *Desalination* **2016**, *387*, 14–24. [CrossRef]
11. Yin, J.; Kim, E.S.; Yang, J.; Deng, B. Fabrication of a novel thin-film nanocomposite (TFN) membrane containing MCM-41 silica nanoparticles (NPs) for water purification. *J. Membr. Sci.* **2012**, *423*, 238–246. [CrossRef]

12. Yurekli, Y. Removal of heavy metals in wastewater by using zeolite nano-particles impregnated polysulfone membranes. *J. Hazard. Mater.* **2016**, *309*, 53–64. [CrossRef]
13. Yang, X.; Sun, P.; Zhang, H.; Xia, Z.; Waldman, R.Z.; Mane, A.U.; Elam, J.W.; Shao, L.; Darling, S.B. Polyphenol-sensitized atomic layer deposition for membrane interface hydrophilization. *Adv. Funct. Mater.* **2020**, *30*, 1910062. [CrossRef]
14. Yang, H.C.; Xie, Y.; Chan, H.; Narayanan, B.; Chen, L.; Waldman, R.Z.; Sankaranarayanan, S.K.; Elam, J.W.; Darling, S.B. Crude-oil-repellent membranes by atomic layer deposition: Oxide interface engineering. *ACS Nano* **2018**, *12*, 8678–8685. [CrossRef] [PubMed]
15. Zhang, H.; Mane, A.U.; Yang, X.; Xia, Z.; Barry, E.F.; Luo, J.; Wan, Y.; Elam, J.W.; Darling, S.B. Visible-Light-Activated Photocatalytic Films toward Self-Cleaning Membranes. *Adv. Funct. Mater.* **2020**, *30*, 2002847. [CrossRef]
16. Lee, A.; Libera, J.A.; Waldman, R.Z.; Ahmed, A.; Avila, J.R.; Elam, J.W.; Darling, S.B. Conformal nitrogen-doped TiO_2 photocatalytic coatings for sunlight-activated membranes. *Adv. Sustain. Syst.* **2017**, *1*, 1600041. [CrossRef]
17. Zaferani, S.H. Introduction of polymer-based nanocomposites. In *Polymer-Based Nanocomposites for Energy and Environmental Applications*; Elsevier: Amsterdam, The Netherlands, 2018; pp. 1–25.
18. Jamróz, E.; Kulawik, P.; Kopel, P. The effect of nanofillers on the functional properties of biopolymer-based films: A review. *Polymers* **2019**, *11*, 675. [CrossRef] [PubMed]
19. Pourjafar, S.; Rahimpour, A.; Jahanshahi, M. Synthesis and characterization of PVA/PES thin film composite nanofiltration membrane modified with TiO_2 nanoparticles for better performance and surface properties. *J. Ind. Eng. Chem.* **2012**, *18*, 1398–1405. [CrossRef]
20. Leong, S.; Razmjou, A.; Wang, K.; Hapgood, K.; Zhang, X.; Wang, H. TiO_2 based photocatalytic membranes: A review. *J. Membr. Sci.* **2014**, *472*, 167–184. [CrossRef]
21. Mozia, S. Photocatalytic membrane reactors (PMRs) in water and wastewater treatment. A review. *Sep. Purif. Technol.* **2010**, *73*, 71–91. [CrossRef]
22. Imam, S.H.; Cinelli, P.; Gordon, S.H.; Chiellini, E. Characterization of biodegradable composite films prepared from blends of poly (vinyl alcohol), cornstarch, and lignocellulosic fiber. *J. Polym. Environ.* **2005**, *13*, 47–55. [CrossRef]
23. Nor, N.A.M.; Jaafar, J.; Ismail, A.F.; Mohamed, M.A.; Rahman, M.A.; Othman, M.H.D.; Lau, W.J.; Yusof, N. Preparation and performance of PVDF-based nanocomposite membrane consisting of TiO_2 nanofibers for organic pollutant decomposition in wastewater under UV irradiation. *Desalination* **2016**, *391*, 89–97. [CrossRef]
24. Jaleh, B.; Etivand, E.S.; Mohazzab, B.F.; Nasrollahzadeh, M.; Varma, R.S. Improving wettability: Deposition of TiO_2 nanoparticles on the O_2 plasma activated polypropylene membrane. *Int. J. Mol. Sci.* **2019**, *20*, 3309. [CrossRef] [PubMed]
25. Aslam, M.; Kalyar, M.A.; Raza, Z.A. Polyvinyl alcohol: A review of research status and use of polyvinyl alcohol based nanocomposites. *Polym. Eng. Sci.* **2018**, *58*, 2119–2132. [CrossRef]
26. Montallana, A.D.S.; Lai, B.Z.; Chu, J.P.; Vasquez, M.R., Jr. Enhancement of photodegradation efficiency of PVA/TiO_2 nanofiber composites via plasma treatment. *Mater. Today Commun.* **2020**, *24*, 01183. [CrossRef]
27. Lou, L.; Kendall, R.J.; Ramkumar, S. Comparison of Hydrophilic PVA/TiO_2 and Hydrophobic PVDF/TiO_2 Microfiber Webs on the Dye Pollutant Photo-catalyzation. *J. Environ. Chem. Eng.* **2020**, 103914. [CrossRef]
28. Bolto, B.; Tran, T.; Hoang, M.; Xie, Z. Crosslinked poly (vinyl alcohol) membranes. *Progress Polym. Sci.* **2009**, *34*, 969–981. [CrossRef]
29. Ngo, T.H.A.; Nguyen, D.T.; Do, K.D.; Nguyen, T.T.M.; Mori, S.; Tran, D.T. Surface modification of polyamide thin film composite membrane by coating of titanium dioxide nanoparticles. *J. Sci. Adv. Mater. Device* **2016**, *1*, 468–475. [CrossRef]
30. Sakarkar, S.; Muthukumaran, S.; Jegatheesan, V. Evaluation of polyvinyl alcohol (PVA) loading in the PVA/titanium dioxide (TiO_2) thin film coating on polyvinylidene fluoride (PVDF) membrane for the removal of textile dyes. *Chemosphere* **2020**, *257*, 127144. [CrossRef] [PubMed]
31. Li, X.; Chen, Y.; Hu, X.; Zhang, Y.; Hu, L. Desalination of dye solution utilizing PVA/PVDF hollow fiber composite membrane modified with TiO_2 nanoparticles. *J. Membr. Sci.* **2014**, *471*, 118–129. [CrossRef]
32. Liu, X.; Chen, Q.; Lv, L.; Feng, X.; Meng, X. Preparation of transparent PVA/TiO_2 nanocomposite films with enhanced visible-light photocatalytic activity. *Catal. Commun.* **2015**, *58*, 30–33. [CrossRef]
33. Yang, H.; Zhang, J.; Song, Y.; Xu, S.; Jiang, L.; Dan, Y. Visible light photocatalytic activity of C-PVA/TiO_2 composites for degrading rhodamine B. *Appl. Surface Sci.* **2015**, *324*, 645–651. [CrossRef]
34. Sakarkar, S.; Muthukumaran, S.; Jegatheesan, V. Polyvinylidene Fluoride and Titanium Dioxide Ultrafiltration Photocatalytic Membrane: Fabrication, Morphology, and Its Application in Textile Wastewater Treatment. *J. Environ. Eng.* **2020**, *146*, 04020053. [CrossRef]
35. Goncalves, M.S.; Oliveira-Campos, A.M.; Pinto, E.M.; Plasencia, P.M.; Queiroz, M.J.R. Photochemical treatment of solutions of azo dyes containing TiO_2. *Chemosphere* **1999**, *39*, 781–786. [CrossRef]
36. Jye, L.W.; Ismail, A.F. *Nanofiltration Membranes: Synthesis, Characterization and Applications*; CRC Press: Boca Raton, FL, USA, 2016.
37. Mänttäri, M.; Pihlajamäki, A.; Nyström, M. Effect of pH on hydrophilicity and charge and their effect on the filtration efficiency of NF membranes at different pH. *J. Membr. Sci.* **2006**, *280*, 311–320. [CrossRef]
38. Madaeni, S.S.; Zinadini, S.; Vatanpour, V. A new approach to improve antifouling property of PVDF membrane using in situ polymerization of PAA functionalized TiO_2 nanoparticles. *J. Membr. Sci.* **2011**, *380*, 155–162. [CrossRef]

39. Kaler, V.; Pandel, U.; Duchaniya, R.K. Development of TiO_2/PVA nanocomposites for application in solar cells. *Mater. Today Proc.* **2018**, *5*, 6279–6287. [CrossRef]
40. Cruz, N.K.O.; Semblante, G.U.; Senoro, D.B.; You, S.J.; Lu, S.C. Dye degradation and antifouling properties of polyvinylidene fluoride/titanium oxide membrane prepared by sol–gel method. *J. Taiwan Inst. Chem. Eng.* **2014**, *45*, 192–201. [CrossRef]
41. Gaikwad, V.L.; Choudhari, P.B.; Bhatia, N.M.; Bhatia, M.S. Characterization of pharmaceutical nanocarriers: In vitro and in vivo studies. In *Nanomaterials for Drug Delivery and Therapy*; William Andrew Publishing: Norwich, NY, USA, 2019; pp. 33–58.
42. Guo, J.; Farid, M.U.; Lee, E.J.; Yan, D.Y.S.; Jeong, S.; An, A.K. Fouling behavior of negatively charged PVDF membrane in membrane distillation for removal of antibiotics from wastewater. *J. Membr. Sci.* **2018**, *551*, 12–19. [CrossRef]
43. Wei, Y.; Chu, H.Q.; Dong, B.Z.; Li, X.; Xia, S.J.; Qiang, Z.M. Effect of TiO_2 nanowire addition on PVDF ultrafiltration membrane performance. *Desalination* **2011**, *272*, 90–97. [CrossRef]
44. Shi, F.; Ma, Y.; Ma, J.; Wang, P.; Sun, W. Preparation and characterization of PVDF/TiO_2 hybrid membranes with different dosage of nano-TiO_2. *J. Membr. Sci.* **2012**, *389*, 522–531. [CrossRef]
45. Aziz, S.B.; Abdulwahid, R.T.; Rasheed, M.A.; Abdullah, O.G.; Ahmed, H.M. Polymer blending as a novel approach for tuning the SPR peaks of silver nanoparticles. *Polymers* **2017**, *9*, 486. [CrossRef]
46. Liu, H.; Liang, Y.; Hu, H.; Wang, M. Hydrothermal synthesis of mesostructured nanocrystalline TiO_2 in an ionic liquid–water mixture and its photocatalytic performance. *Solid State Sci.* **2009**, *11*, 1655–1660. [CrossRef]
47. Yu, X.; Mi, X.; He, Z.; Meng, M.; Li, H.; Yan, Y. Fouling resistant CA/PVA/TiO_2 imprinted membranes for selective recognition and separation salicylic acid from waste water. *Front. Chem.* **2017**, *5*, 2. [CrossRef]
48. Hidalgo, A.; Gómez, M.; Murcia, M.; Serrano, M.; Rodriguez-Schmidt, R.; Escudero, P. Behaviour of polysulfone ultrafiltration membrane for dyes removal. *Water Sci. Technol.* **2018**, *77*, 2093–2100. [CrossRef] [PubMed]
49. Nakabayashi, Y.; Nosaka, Y. The pH dependence of OH radical formation in photo-electrochemical water oxidation with rutile TiO_2 single crystals. *Phys. Chem. Chem. Phys.* **2015**, *17*, 30570–30576. [CrossRef]
50. Kiwaan, H.A.; Atwee, T.M.; Azab, E.A.; El-Bindary, A.A. Photocatalytic degradation of organic dyes in the presence of nanostructured titanium dioxide. *J. Mol. Struct.* **2020**, *1200*, 127115. [CrossRef]
51. Rajaeian, B.; Heitz, A.; Tade, M.O.; Liu, S. Improved separation and antifouling performance of PVA thin film nanocomposite membranes incorporated with carboxylated TiO_2 nanoparticles. *J. Membr. Sci.* **2015**, *485*, 48–59. [CrossRef]
52. Wang, Q.; Wang, Z.; Zhang, J.; Wang, J.; Wu, Z. Antifouling behaviours of PVDF/nano-TiO_2 composite membranes revealed by surface energetics and quartz crystal microbalance monitoring. *RSC Adv.* **2014**, *4*, 43590–43598. [CrossRef]
53. Riaz, S.; Park, S.J. An overview of TiO_2-based photocatalytic membrane reactors for water and wastewater treatments. *J. Ind. Eng. Chem.* **2020**, *84*, 23–41. [CrossRef]
54. Khataee, A.R.; Kasiri, M.B. Photocatalytic degradation of organic dyes in the presence of nanostructured titanium dioxide: Influence of the chemical structure of dyes. *J. Mol. Catal. A Chem.* **2010**, *328*, 8–26. [CrossRef]
55. Reinosa, J.J.; Docio, C.M.Á.; Ramírez, V.Z.; Lozano, J.F.F. Hierarchical nano ZnO-micro TiO_2 composites: High UV protection yield lowering photodegradation in sunscreens. *Ceram. Int.* **2018**, *44*, 2827–2834. [CrossRef]
56. Deng, F.; Zou, J.P.; Zhao, L.N.; Zhou, G.; Luo, X.B.; Luo, S.L. Nanomaterial-based photocatalytic hydrogen production. In *Nanomaterials for the Removal of Pollutants and Resource Reutilization*; Elsevier: Amsterdam, The Netherlands, 2019; pp. 59–82. [CrossRef]
57. Chin, S.S.; Chiang, K.; Fane, A.G. The stability of polymeric membranes in a TiO_2 photocatalysis process. *J. Membr. Sci.* **2006**, *275*, 202–211. [CrossRef]

Article

Enhancing the Antibacterial Properties of PVDF Membrane by Hydrophilic Surface Modification Using Titanium Dioxide and Silver Nanoparticles

Kajeephan Samree [1], Pen-umpai Srithai [1], Panaya Kotchaplai [2], Pumis Thuptimdang [3,4], Pisut Painmanakul [5,6,7], Mali Hunsom [8] and Sermpong Sairiam [1,*]

1. Department of Environmental Science, Faculty of Science, Chulalongkorn University, Bangkok 10330, Thailand; kajeephan0603@gmail.com (K.S.); s.penumpai@gmail.com (P.-u.S.)
2. Institute of Biotechnology and Genetic Engineering, Chulalongkorn University, Bangkok 10330, Thailand; p.kotchaplai@gmail.com
3. Department of Chemistry, Faculty of Science, Chiang Mai University, Chiang Mai 50200, Thailand; pumis.th@gmail.com
4. Environmental Science Research Center, Faculty of Science, Chiang Mai University, Chiang Mai 50200, Thailand
5. Department of Environmental Engineering, Faculty of Engineer, Chulalongkorn University, Bangkok 10300, Thailand; pisut114@hotmail.com
6. Research Program on Development of Technology and Management Guideline for Green Community, Center of Excellence on Hazardous Substance Management (HSM), Bangkok 10330, Thailand
7. Research Unit on Technology for Oil Spill and Contamination Management, Chulalongkorn University, Bangkok 10330, Thailand
8. Academy of Science, The Royal Society of Thailand, Office of the Royal Society, Dusit, Bangkok 10300, Thailand; mhunsom@gmail.com
* Correspondence: sermpong.s@chula.ac.th

Received: 21 July 2020; Accepted: 14 October 2020; Published: 15 October 2020

Abstract: This work investigates polyvinylidene fluoride (PVDF) membrane modification to enhance its hydrophilicity and antibacterial properties. PVDF membranes were coated with nanoparticles of titanium dioxide (TiO_2-NP) and silver (AgNP) at different concentrations and coating times and characterized for their porosity, morphology, chemical functional groups and composition changes. The results showed the successfully modified PVDF membranes containing TiO_2-NP and AgNP on their surfaces. When the coating time was increased from 8 to 24 h, the compositions of Ti and Ag of the modified membranes were increased from 1.39 ± 0.13 to 4.29 ± 0.16 and from 1.03 ± 0.07 to 3.62 ± 0.08, respectively. The water contact angle of the membranes was decreased with increasing the coating time and TiO_2-NP/AgNP ratio. The surface roughness and permeate fluxes of coated membranes were increased due to increased hydrophilicity. Antimicrobial and antifouling properties were investigated by the reduction of *Escherichia coli* cells and the inhibition of biofilm formation on the membrane surface, respectively. Compared with that of the original PVDF membrane, the modified membranes exhibited antibacterial efficiency up to 94% against *E. coli* cells and inhibition up to 65% of the biofilm mass reduction. The findings showed hydrophilic improvement and an antimicrobial property for possible wastewater treatment without facing the eminent problem of biofouling.

Keywords: polyvinylidene fluoride membrane (PVDF); titanium dioxide nanoparticles (TiO_2-NP); silver nanoparticles (AgNP); antibacterial property; antifouling property

1. Introduction

Nowadays, the number of industries is rapidly increasing, thereby posing the risk of serious water pollution. Membrane filtration is among the most popular methods for sustainable wastewater treatment because of its advantages including no phase changes or chemical addition, simple operation, and relatively low energy consumption [1]. Normally, membranes are typically fabricated from hydrophobic materials such as polyvinylidene fluoride (PVDF), polytetrafluoroethylene (PTFE), polyethylene (PE), or polypropylene (PP). Of all the materials used, PVDF is generally applied as a membrane material due to its thermal stability, excellent chemical resistance, and good membrane-forming ability [2]. However, PVDF is a semi-crystalline polymer with repeating units of $-CH_2-CF_2-$, which form hydrophobic structures that can make the membrane more prone to fouling [3–5]. Hydrophobic PVDF membrane is susceptible to fouling when the contacting aqueous solution contains hydrophobic species such as protein, resulting in easy absorption on the membrane surface or blockage of the membrane pores leading to decreased permeability [6,7]. Membrane biofouling involves the attachment of microorganisms on the membrane surface to form biofilm, a cluster of cells and their produced extracellular polymeric substances (EPS), which can significantly decrease the separation performance of the membrane [1]. Biofilm and its associated EPS are mainly responsible for the membrane water flux decline since their coverage on the membrane surfaces can block membrane pores and increase water transport resistance [8]. This can shorten the working life of the membrane, increase the operation cost, and finally restrict successful applications. To solve this problem, the membrane hydrophilicity must be increased to induce fouling resistance, thereby preventing the attachment of microbial cells along with the adsorption and deposition of hydrophobic pollutants onto the membrane surface [9].

There have been many attempts at the modification of the hydrophilic layer on the existing membrane surface in order to enhance the hydrophilicity, antibacterial properties and membrane performance, resulting in more efficient membrane applications such as water and wastewater treatment [10–12]. One of the most common and effective techniques to increase the hydrophilicity on the membrane surface is the addition of inorganic nano-/micro-particles including SiO_2, TiO_2, Al_2O_3, and ZrO_2, which contain an abundance of polar groups on their particle surfaces [12,13]. One of the most popular inorganic particles used is titanium dioxide nanoparticles (TiO_2-NP) because of its non-toxicity, low cost, and superhydrophilicity [14]. Furthermore, the catalytic activity of TiO_2-NP was proved to be capable of killing a wide range of microorganisms including endospores, as well as fungi, algae, protozoa, and viruses [15]. For decades, silver nanoparticles (AgNP) have been generally used as an effective broad-spectrum biocide to functionalize filtration membrane surfaces for biofouling mitigation [16,17]. The antibacterial mechanism of AgNP, as reported in previous literature, is the release of silver ions (Ag^+) that can induce cell membrane damage, promote generation of reactive oxygen species (ROS) and disrupt adenosine triphosphate (ATP) production and DNA replication, ultimately causing the death of bacteria [18]. For the purpose of enhancing the membrane fouling, the changes of membrane properties during the membrane preparation process and surface membrane modification were mainly employed. A previous study has shown that AgNPs deposited on PVDF membrane surface by physical vapor deposition could enhance the antibacterial properties and decrease bacteria growth [19]. However, the modification of PVDF membrane by TiO_2-NP and AgNP using the dipped coating technique still suffers from a research gap requiring improvement. Also, the contamination of *Escherichia coli* still poses serious public health issues and numerous economic losses [20]. Therefore, TiO_2-NP/AgNP-modified PVDF membrane could be useful for the treatment of wastewater contaminated with *E. coli*. By using this modification method, the membrane performance could be improved for long term use, thereby offering a more sustainable wastewater treatment technology.

This study aims to enhance the antibacterial and antifouling properties of the PVDF membrane modified by TiO_2-NP and AgNP using the dipped coating method. The modified membrane was characterized by scanning electron microscopy (SEM) with energy dispersive X-ray spectroscopy (EDS) before being identified for its morphology, contact angle, and porosity. By using *E. coli* as a model strain, the antibacterial property and biofilm formation inhibition of the modified membranes were investigated.

2. Methodology

2.1. Chemicals and Materials

PVDF flat sheet membrane was purchased from ANOW® (ANOW®, Hangzhou, China) with the membrane specifications already reported by the manufacturer (Table 1). A commercial TiO_2-NP and AgNP were supplied from Prime Nanotechnology (Bangkok, Thailand). The properties of TiO_2-NP and AgNP as obtained from the manufacturer are summarized in Table 2. Deionized (DI) water was used for solution preparation.

Table 1. Specifications of the polyvinylidene fluoride (PVDF) membrane.

Pore Size (µm)	0.45
Membrane Porosity (%)	80
Thickness (µm)	110
Color	White

Table 2. Properties of titanium dioxide and silver nanoparticles (TiO_2-NP and AgNP).

Properties	TiO_2-NP	AgNP
Appearance	White Powder	Yellow Brown Colloid
Crystalline Structure	80% Anatase, 20% Rutile	-
Primary Particle Size (nm)	21	-
Average Particle Size (nm)	-	5–20
Particle Shape	-	Nanospheres
pH	-	6 ± 1
Specific gravity (g mL^{-1})	-	1.01
Tamped density (g L^{-1})	130	-
Specific surface area (m^2 g^{-1})	50	-
Content (wt%)	>99.5	-

2.2. Experimental Procedures

Membrane Modification via Titanium Dioxide and Silver Nanoparticles (TiO_2-NP and AgNP)

First, TiO_2-NP powders were dispersed in 1 L of deionized (DI) water and sonicated for 15 min using an ultrasonic bath (DT100SH, Bandelin, Berlin, Germany) at 320 W, 35 kHz to obtain a homogeneous TiO_2-NP suspension. Second, AgNP was then added into the solution and the mixture of solution was stirred by magnetic stirrer for 3 min. For membrane modification (Figure 1), 5 pieces of PVDF flat sheet membranes (2 × 7.5 cm^2) were coated by dipping into the 1 L of nanoparticle suspension (75 cm^2/L) The coating times (0–24 h) and concentrations of TiO_2-NP and AgNP (0 and 10 ppm) were varied as shown in Table 3. After that, the modified membranes were incubated in an oven (ULM700, Memmert, Schwabach, Germany) at 60 °C for 1 h to eliminate excess liquid and kept in a desiccator to completely remove moisture before being used for analysis and further experiments.

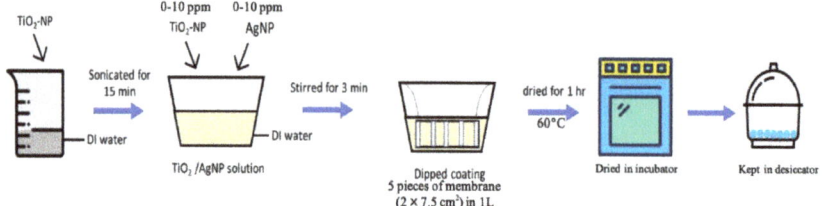

Figure 1. Schematic diagram for TiO$_2$-NP/AgNP dipped coating of 75 cm^2 of flat sheet PVDF membrane into 1L solution.

Table 3. Concentrations of TiO$_2$-NP and AgNP and coating time for membrane modification.

Membrane	Concentration of TiO$_2$-NP (ppm)	Concentration of AgNP (ppm)	Immersion Time (h)
Original PVDF			
M1	0	0	0
Control PVDF			
TiO$_2$-NP	10	0	24
AgNP	0	10	24
Modified PVDF			
M2	10	10	8
M3	10	10	16
M4	10	10	24
M5	10	20	24
M6	20	10	24

2.3. Membrane Characterization

2.3.1. Contact Angle Measurement

The membrane hydrophilicity was evaluated from the surface contact angle (OCA40, Dataphysics, Filderstadt, Germany). The water contact angle (WCA) of the membrane was measured by the sessile drop method at room temperature. One membrane sample was used for the measurement of each type of membrane (Table 3). DI water was dropped on the top surface of a dried membrane at three different positions on the single piece of membrane, and then the data were presented as a mean of the contact angles with the errors represented as standard deviations.

2.3.2. Morphology and Chemical Composition

Scanning electron microscopy (SEM, JSM-IT500HR, JEOL, Tokyo, Japan) and energy-dispersive X-ray spectroscopy (EDS, JEOL, Tokyo, Japan) were used to investigate the presence of TiO$_2$-NP and AgNP as well as the surface morphology of the membrane. For each type of membrane, one membrane sample was used for the analysis. The membranes were cut into pieces and coated with gold under vacuum condition before the observation to avoid the electrostatic charging through a sputter-coater (Balzers UNION Limited, SCD040, Balzers, Liechtenstein). The chemical compositions of the membranes were analyzed three times on the same membrane sample, and the data obtained were presented as a mean with the errors represented as standard deviations. Besides, the outer surface topographies and roughness of the original and modified PVDF membranes were further observed by atomic force microscopy (AFM, SPA 400, SEIKO, Chiba, Japan). The scan area of 10 μm × 10 μm was reported using a tapping mode.

2.3.3. Porosity Analysis

Porosity of the membrane was evaluated using the gravimetric method. Each sample of the membrane (1.5 × 1.5 cm^2) was weighed using a digital weight balance (MS204S/01, Mettler Toledo,

Greifensee, Switzerland) and recorded as the dry weight (m_2). After that, the membrane was immersed in DI water for 48 h. Then, the sample was weighed immediately and recorded as the wet weight (m_1). The membrane porosity (ε) was calculated using Equation (1) as follows [21]:

$$\varepsilon = \left(\frac{(m_1 - m_2)/\rho_{water}}{\frac{(m_1 - m_2)}{\rho_{water}} + \frac{m_2}{\rho_p}} \right) \times 100\% \qquad (1)$$

where m_1 is the wet membrane weight (g), m_2 is the dry membrane weight (g), ρ_{water} is the water density (0.998 g cm^{-3}), and ρ_p is the PVDF density (1.74 g cm^{-3}). The data of the membrane porosity were presented as a mean with standard deviations calculated from three data of three separated membranes.

2.3.4. Permeation Performances

The permeation performances of both original and modified PVDF flat sheet membranes were performed by an in-house dead end filtration device (Sterlitech Corporation, CF042D, Kent, WA, USA). The effective membrane area of a filtration device is 47 cm^2. For each membrane, the membranes were immersed in DI water for 24 h at room temperature. Prior to measurement, the membrane was pre-compacted at 4 bar for 30 min to obtain the initial flux. DI water was fed into the filtration device at the desired pressure, then, the permeate water obtained at different time intervals was measured using an electronic weighting balance to evaluate the permeate flux (J) which was expressed as Equation (2):

$$J = \frac{V}{A \times t} \qquad (2)$$

where V is the volume of permeate water (L), A is the effective area (m^2) and t is the running time (h)

2.3.5. Functional Group Analysis

Raman spectroscopy was carried out in a DXR Raman microscope (Thermo Fisher Scientific Inc., Madison, WI, USA) using a laser excitation wavelength of 780 nm of laser power. The experiments were operated through an aperture of 50-micron slit and a 10×-objective lens with a laser spot of 3.1 µm. Raman spectra were obtained using a 2 s exposure time with 32 accumulations.

2.4. Antibacterial Test

2.4.1. E. Coli Strains and Growth Condition

E. coli were obtained from multiple tube fermentation technique by the standard protocol of the most probable number of coliform organism test (MPN) for gas and acid productions [22]. First, three different volumes of water sample (10, 1 and 0.1 mL) were transferred into each of five Durham tubes containing 10 mL of lactose broth (LB) and incubated at 35 °C for 48 h (ULM 600, Memmert, Schwabach, Germany). The solution from the test tubes showing gas production was transferred to a new tube containing brilliant green lactose bile broth (BGLB) and incubated overnight at 35 °C. Then, 0.1 mL of the BGLB was spread on nutrient agar (NA) plate before incubated at 37 °C for 24 h. The isolated colonies, identified as *E. coli*, were further restreaked onto an NA plate until a single colony is obtained. Once purified, the isolates were maintained in liquid nutrient broth (NB) at 4 °C. This suspension was used for antibacterial and biofilm inhibition tests.

2.4.2. Antibacterial Test

The experiment was modified from Liu, Yao, Ren, Zhao, and Yuan (2018) [1]. First, the membrane (1.5 × 1.5 cm^2) was immersed into the suspension of *E. coli* cells (pre-adjusted with liquid NB to obtain the initial concentration of 10^6 CFU/mL) with NB in 24 well-plates. Then, the well plates were incubated at 37 °C for 48 h. The membranes were rinsed gently with 0.85% NaCl twice to remove the excess

E. coli suspensions and non-adhered cells. The membranes were then transferred into sterile test tubes containing 10 mL of 0.85% NaCl before vortex mixing for 1 min to detach the cells. Finally, 0.1 mL of cell suspension was spread on NA plate and incubated overnight at 37 °C. The number of viable cells on the membrane was expressed as colony forming unit (CFU/mL). Data were obtained from three wells, each of which contained a single piece of membrane, and represented as a mean and standard deviations.

2.4.3. Biofilm Inhibition Test

The biofilm inhibition test was conducted to determine the performance of the modified membranes to inhibit biofilm formation on surface by measuring the biomass of biofilm. The pre-weighed membranes (1.5 × 1.5 cm^2) were immersed into the suspension of E. coli cells in a similar manner to the antibacterial test, but the incubation time was increased to 72 h to allow biofilm formation. The membranes were rinsed with DI water to remove the loosely bound cells. The membranes were oven-dried at 60 °C for 1 h followed by air-drying. The weight of the membrane was measured by a digital weight balance (UMX2, Mettler Toledo, Greifensee, Switzerland). Biofilm mass was determined by the difference between the weight of the membrane before and after being immersed into the suspension of E. coli cells [23]. Data were obtained from three different pieces of membrane, and represented as a mean and standard deviations.

3. Results and Discussion

3.1. Hydrophilic Membrane Modification

3.1.1. Water Contact Angle (WCA)

(a) Effect of coating time

WCA was measured to evaluate the membrane hydrophilicity that plays a significant role in membrane permeability [24]. The results showed that the WCAs of the original PVDF membrane M1 and the modified membrane M2, M3, and M4 were 84.0° ± 2.1°, 86.3° ± 1.7°, 86.7° ± 0.5° and 35.2° ± 1.6°, respectively (Figure 2). The WCA of M4 was decreased to be lower than that of M2 and M3 due to higher TiO$_2$-NP and AgNP content. This could result from the longest coating time for M4, which allows more TiO$_2$-NP/AgNP to attach on the membrane surface as shown in SEM results in the previous section (Figure 2). Furthermore, polyetherethersulfone incorporated with AgNP exhibited the increased hydrophilicity, leading to the increase of water permeability associated with higher affinity [25]. This was because of high specific surface area and fraction of nanoparticles that provide specific functionality to the modified membrane for facilitating the hydrophilicity and permeability. Therefore, the results indicated that the coating time of 24 h can improve the membrane hydrophilicity.

(b) Effect of TiO$_2$-NP/AgNP concentration

The results of WCAs of the original membrane (M1) and TiO$_2$-NP/AgNP modified membranes (M4–M6) by varied nanoparticle concentrations were illustrated in Figure 2. The WCA of M1 was 84.0° ± 2.1° due to the hydrophobic property of the original membrane. As TiO$_2$-NP/AgNP were introduced into the membranes, the WCA were decreased to 35.2° ± 1.6° (M4), 35.3° ± 1.4° (M5) and 70.4° ± 3.0° (M6) for the nanoparticle ratios of 10 ppm TiO$_2$-NP/10 ppm AgNP, 10 ppm TiO$_2$-NP/20 ppm AgNP and 20 ppm TiO$_2$-NP/10 ppm AgNP, respectively, implying more hydrophilicity. Figure 2 also shows that the WCA of AgNP membrane (control) was decreased to 74.0° ± 0.6°. Since AgNP has high specific surface area and high fraction of surface atom, the adhesion of AgNP on the surface can alter the membrane properties to be more hydrophilic and permeable [25]. On the other hand, the WCA of TiO$_2$-NP membrane (control) was increased to 120.4° ± 1.9°. The differences in contact angle between the control TiO$_2$-NP and AgNP membranes suggested that the optimal ratio between TiO$_2$-NP and AgNP contents could play an important role in the coating and differentiation of WCA. As the contents of TiO$_2$-NP/AgNP were increased to 10 ppm TiO$_2$-NP/20 ppm AgNP and 20 ppm

TiO$_2$-NP/10 ppm AgNP for M4 and M5, their WCAs were similarly decreased. This could be due to the agglomeration of TiO$_2$-NP and AgNP from high concentrations; therefore, the WCAs were not decreased even though the AgNP concentration was increased. Moreover, M6 was coated by highest concentration of TiO$_2$-NP compared to M4 and M5, but the WCA was slightly decreased compared to the original membrane M1 (Figure 2), which could also result from the excess of TiO$_2$-NP on the membrane to cause TiO$_2$-NP agglomeration. According to the results, increasing TiO$_2$-NP and AgNP concentrations did not decrease the WCA; therefore, using 10 ppm TiO$_2$-NP/10 ppm AgNP for M4 coating was adequate to improve the hydrophilicity of the membrane.

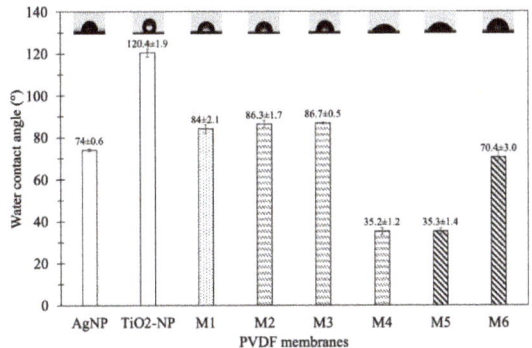

Figure 2. The water contact angle (WCA) of M1 (original), M2 (8 h), M3 (16 h), M4 (24 h) at mixed 10 ppm TiO$_2$-NP/10 ppm AgNP, and M5 (10 ppm TiO$_2$-NP/20 ppm AgNP) and M6 (20 ppm TiO$_2$-NP/10 ppm AgNP) at coating time of 24 h.

3.1.2. Membrane Morphology

(a) Effect of coating time

The PVDF membrane was immersed in the solution of 10 ppm TiO$_2$-NP and 10 ppm AgNP for varied periods of time to study the effect of coating time on the membrane. The results showed that, compared with the original membrane (Figure 3a), the TiO$_2$-NP/AgNP was observed on the modified membranes (Figure 3b–d), suggesting the successful coating. More particles of TiO$_2$-NP/AgNP were found on the membrane surface when the coating time was increased from 8 to 24 h for M2 (Figure 3b) and M4 (Figure 3d), respectively. Figure 3b shows that the agglomerated TiO$_2$-NP was distributed throughout the surface while AgNP was hardly spotted; however, as the coating time increased, TiO$_2$-NP and AgNP were both clearly dispersed and observed on membrane M3 (Figure 3c) and M4 (Figure 3d). In addition, TiO$_2$-NP and AgNP were mostly attached together because these nanoparticles have high surface energy that makes them tend to agglomerate to reach more stable state [26]. EDS mapping of TiO$_2$-NP and AgNP distributions on M4 were also carried out and presented in Figure 4. It can be seen clearly that the distributions of TiO$_2$-NP and AgNP were observed on the membrane surface after coating of 10 ppm TiO$_2$-NP and 10 ppm AgNP for 24 h as illustrated in Figure 4a–c. These results supported the successful coating and the layer of TiO$_2$-NP and AgNP could improve the hydrophilicity and possibly increase the membrane permeate flux.

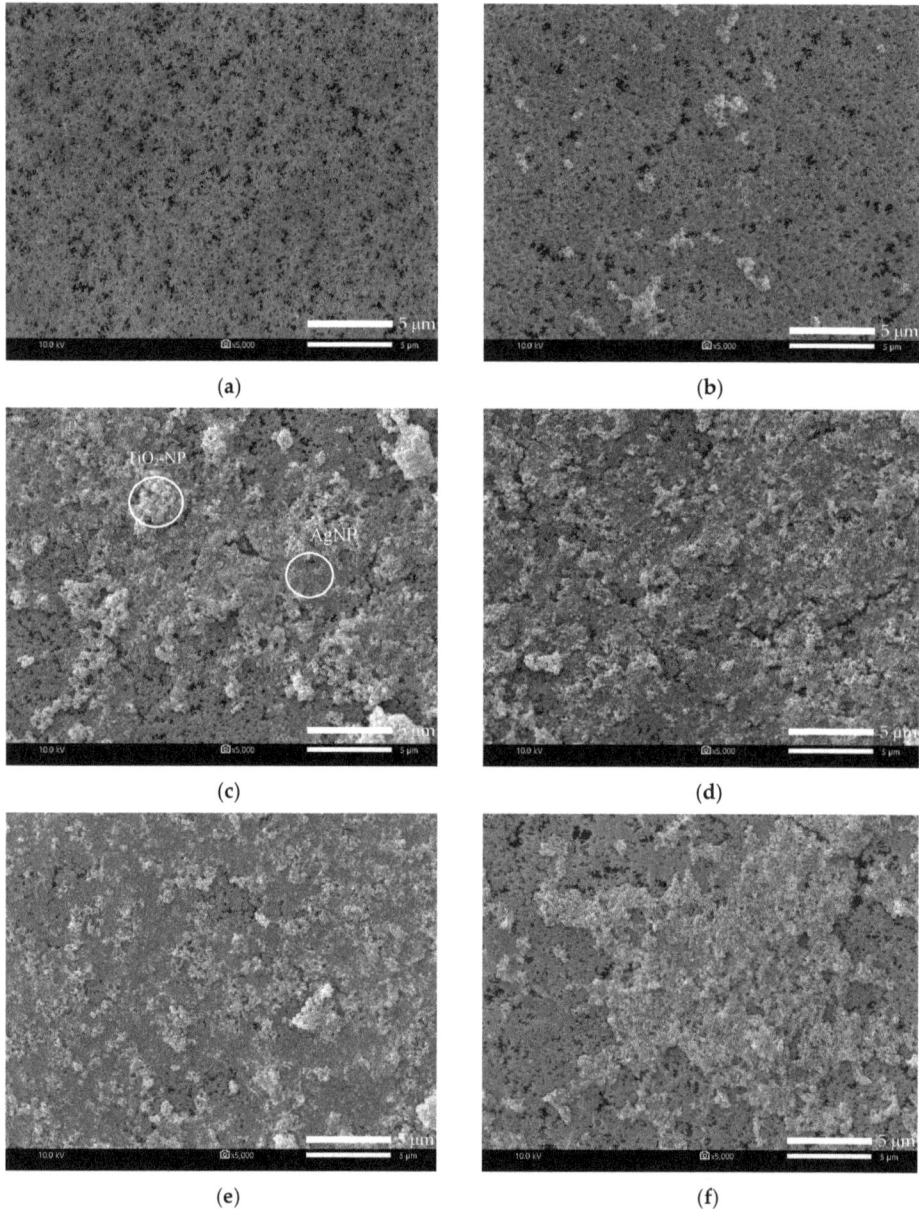

Figure 3. Scanning electron microscope (SEM) images of the membranes modified by different coating times: (**a**) M1 (original); (**b**) M2 (8 h); (**c**) M3 (16 h); (**d**) M4 (24 h) at mixed 10 ppm TiO_2-NP/10 ppm AgNP; (**e**) M5 (10 ppm TiO_2-NP/20 ppm AgNP); and (**f**) M6 (20 ppm TiO_2-NP/10 ppm AgNP) at coating time of 24 h (magnification ×5000).

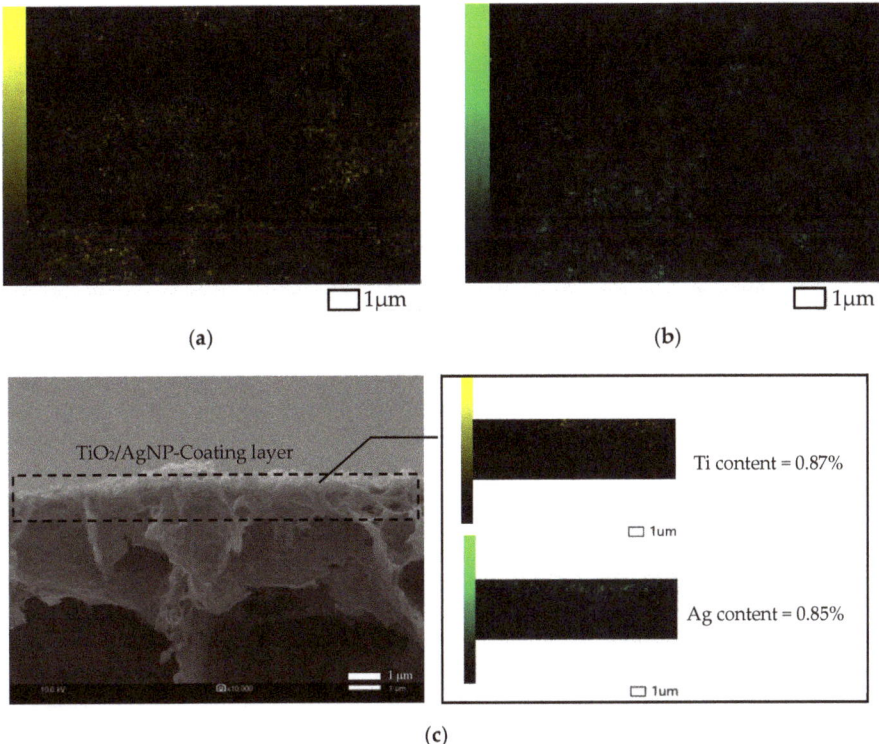

Figure 4. Mapping of particle distribution of modified membrane with 10 ppm TiO$_2$ and 10 ppm AgNP for 24 h (M4): (**a**) TiO$_2$-NP on membrane surface; (**b**) AgNP on membrane surface; and (**c**) TiO$_2$-NP and AgNP of X-section (magnification ×10,000).

Additionally, elemental compositions of the modified membranes were detected by EDS. Three elements including Ti, O, and Ag were analyzed to confirm the existence of TiO$_2$-NP and AgNP from the membrane modification, and the results are shown in Table 4. From the results, the modified membranes M2–M4 contain more Ti than Ag even though the same concentrations of TiO$_2$-NP and AgNP were used, which indicates that Ti might adhere on the membrane surface better than Ag. Furthermore, the percentage of O was higher than that of Ti in all membranes due to the elemental ratio of TiO$_2$. Table 4 also confirms the results in Figure 2 that the number of TiO$_2$/AgNP that adhered on the membrane surface depends on the coating time. Table 4 shows the composition of Ti, O and Ag on M4 which was lower than M3 (coating time of 24 h). This might be due to more agglomeration of TiO$_2$-NP/AgNP on M4 compared with M3, which decreases the surface area of TiO$_2$-NP/AgNP on the membrane. These results indicated that TiO$_2$-NP/AgNP were highly adhered by using the coating time of 16 h (membrane M3).

Table 4. Chemical compositions of the modified membranes. (The standard deviation (S.D.) was obtained from 1 sample of each condition).

Membrane	C (%)	F (%)	O (%)	Ti (%)	Ag (%)
M1	55.11 ± 0.18	43.61 ± 0.22	1.28 ± 0.08	-	-
M2	58.16 ± 0.21	35.05 ± 0.28	4.36 ± 0.14	1.39 ± 0.13	1.03 ± 0.07
M3	35.89 ± 0.17	30.42 ± 0.34	20.51 ± 0.27	8.33 ± 0.22	4.84 ± 0.11
M4	45.02 ± 0.17	34.42 ± 0.27	12.65 ± 0.18	4.29 ± 0.16	3.62 ± 0.08
M4^{after}	55.62 ± 0.07	37. 59 ± 0.11	5. 36 ± 0.06	0.93 ± 0.02	0.51 ± 0.01
M5	46.57 ± 0.18	33.61 ± 0.30	12.19 ± 0.20	4.06 ± 0.16	3.57 ± 0.09
M6	30.87 ± 0.16	25.46 ± 0.31	24.48 ± 0.30	13.80 ± 0.26	5.39 ± 0.11

Remark: (1) M4^{after} is the chemical compositions of membrane after testing with pure water flux. (2) S.D. is varied between different elements and analytical lines and for the same element in different matrices.

(b) Effect of TiO_2-NP/AgNP concentration

The morphology of the modified membranes using different nanoparticle concentrations, which were 10 ppm TiO_2-NP/10 ppm AgNP (M4), 10 ppm TiO_2-NP/20ppm AgNP (M5) and 20 ppm TiO_2-NP/10 ppm AgNP (M6), are shown in Figure 3. The smoother surface of the original PVDF membrane (M1) showed that there are no particles on the surface compared with that of the modified membranes, in which a large number of TiO_2-NP/AgNP can be observed. When the concentrations of TiO_2-NP and AgNP increased, higher distribution of the nanoparticles on the membrane surface can be observed (Figure 3 e–f). Likewise, Yang, Peng, Wang, and Liu (2010) reported that high content of TiO_2 concentration motivates the aggregation phenomenon on the membrane surface [27]. Furthermore, the presence of TiO_2-NP and AgNP on the membranes was confirmed by EDS as reported in Table 4. The results show that the amounts of Ti and O adhered on M6 were twice as much as M4 and M5 due to higher TiO_2-NP concentration (20 ppm) used for M6. From the results of WCA (Figure 2) and the TiO_2-NP and AgNP distribution mappings in Figure 4, it was evident that the WCA was positively correlated to TiO_2-NP and AgNP content on membrane surface. It could be concluded that the deposited TiO_2-NP and AgNP on membrane surface would play important role for enhancing hydrophilicity.

3.1.3. Surface Roughness

(a) Effect of coating time

The AFM images of the membrane coated with the solution of 10 ppm TiO_2-NP and 10 ppm AgNP at different times (8–24 h) are shown in Figure 5. As observed, the original PVDF membrane (M1) exhibits a smooth surface area with the mean surface roughness (R_a) value of 84.7 nm while M2, M3, and M4 membranes show rougher surface at 79.3, 115.7, and 156.0 nm, respectively, with the increase of coating time. The increase of coating time leads to more agglomeration of TiO_2-NP/AgNP particles on the membrane, which could dramatically enhance the surface roughness and, therefore, efficiently increase hydrophilicity. The lower WCA obtained after TiO_2-NP/AgNP coating on membrane surface could be due to the enhance surface roughness. According to the Wenzel's model, the wettability of surface is amplified by increasing the surface roughness; in other words, an increase in surface roughness leads to the increase of surface hydrophilicity [7,28,29].

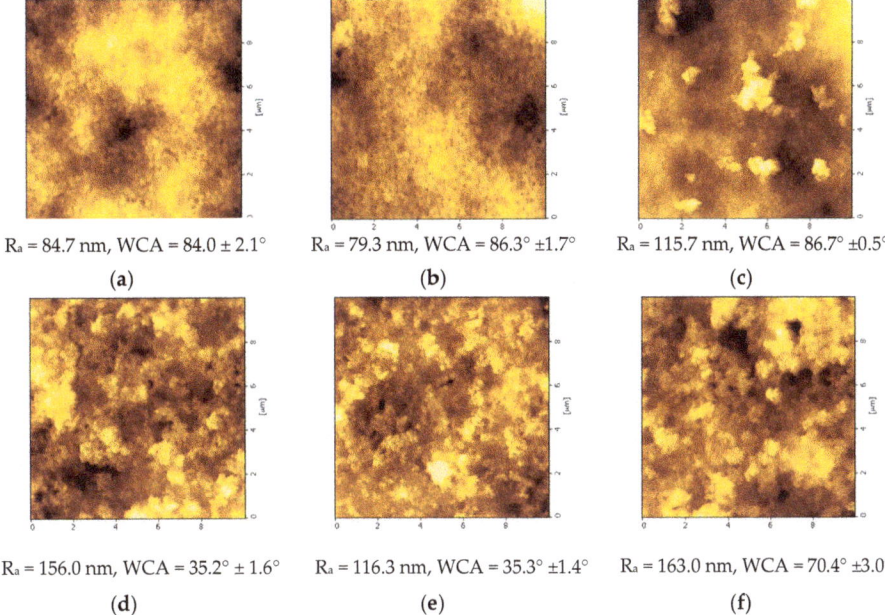

Figure 5. Atomic force microscopy (AFM) image topography of membranes: (**a**) M1 (original); (**b**) M2 (8 h); (**c**) M3 (16 h); (**d**) M4 (24 h) at mixed 10 ppm TiO_2-NP/10 ppm AgNP; (**e**) M5 (10 ppm TiO_2-NP/20 ppm AgNP); and (**f**) M6 (20 ppm TiO_2-NP/10 ppm AgNP) at coating time of 24 h.

(b) Effect of TiO_2-NP/AgNP concentration

According to Figure 5, the surface roughness of PVDF flat sheet membrane was also affected by the addition of TiO_2-NP/AgNP. The AFM images show that the surface of the membranes is rougher compared with the original PVDF membrane like presented in SEM images. R_a of M5 and M6 was 116.3 and 163.0 nm, respectively. R_a of the modified membrane was increased when the TiO_2-NP/AgNP concentration was increased from 10 ppm to 20 ppm. There is a difference of high peak and low valley with increasing the TiO_2-NP/AgNP concentration. The increase of high peaks leads to higher surface roughness. This could explain the lower contact angle of modified membrane compared with the original membrane, showing higher surface roughness.

3.1.4. Porosity

(a) Effect of coating time

The porosities of the original PVDF membrane and the membranes modified by varying coating time are reported in Table 5 and calculated by Equation (1). The results showed that increasing the coating time slightly reduced the porosity of the membrane. The porosity was increased from 61.33% for the original membrane M1 to 67.19%, 67.18%, and 66.74% for M2, M3, and M4, with the coating time of 8 h, 16 h and 24 h, respectively. This could be explained by the fact that longer coating time resulted in more TiO_2-NP and AgNP agglomerations, which increased the opportunity for large particles to block the membrane pores.

Table 5. Porosity of the membranes.

Membrane	Porosity (%)
M1	61.33 ± 2.83
M2	67.19 ± 3.70
M3	67.81 ± 3.79
M4	66.74 ± 2.45
M5	66.45 ± 2.78
M6	64.10 ± 4.73
TiO_2-NP (control)	65.99 ± 10.35
AgNP (control)	66.52 ± 3.35

(b) Effect of TiO_2-NP/AgNP concentration

The porosities of the original PVDF membrane and membranes modified by varying TiO_2-NP/AgNP concentrations were shown in Table 5. The modified membranes with the nanoparticle concentrations of 10 ppm TiO_2-NP/10 ppm AgNP (M4), 10 ppm TiO_2-NP/20 ppm AgNP (M5), and 20 ppm TiO_2-NP/10 ppm AgNP (M6) provided the increased porosities of 66.74% ± 2.45%, 66.45% ± 2.78% and 64.10% ± 4.73%, respectively, compared with 61.33% ± 2.83% of the original PVDF membrane (M1). Previous study has also found that the addition of nanoparticles on the membrane could increase the porosity [30]. However, among the modified membranes, the porosity was decreased when the content of the nanoparticles increased, which might be due to the blockage of membrane pores by the aggregation of nanoparticles [9]. A study found that the addition of 4 wt% TiO_2-NP in combination with PVDF and sulfonated polyethgersulfone (PES) exhibited the reduction of water permeability owing to the decreased porosities [31]. In addition, TiO_2-NP could induce the extremely rapid precipitation and produce a thick dense top layer that can also block the membrane pores [32]. The results indicated that the modified membranes with TiO_2-NP and AgNP were improved in their porosity, whereas increasing the TiO_2-NP/AgNP concentrations did not induce any porosity improvement.

3.1.5. Pure Water Flux of Membranes

Pure water flux of the original and modified PVDF flat sheet membranes was measured at different time intervals under 4 bar of pressure to evaluate the water permeability. Figure 6 exhibits the initial pure water flux ($t = 0$, after setting for 30 min) and the results showed that the modified PVDF membrane coated with TiO_2-NP and AgNP gave higher flux than that of the original PVDF membrane. The original membrane (M1) gave the lowest initial flux at 444.7 L m^{-2} h^{-1} while the M2 and M4 provided the fluxes up to 548.8 and 524.1 L m^{-2} h^{-1}, respectively. The coating time of TiO_2-NP and AgNP affected the pure water flux to a greater extent. The presence of hydroxyl functional group from TiO_2-NP and AgNP contributed to improve the hydrophilicity, thereby improving the pure water flux. M4 had the lowest contact angle (more hydrophilic) but the initial flux was lower than that of M2. This was because the coating time of 24 h (M4) could allow more TiO_2-NP and AgNP accumulation on the membrane surface and membrane pores, resulting in the increased membrane thickness and then lower water flux compared to 8 h (M2).

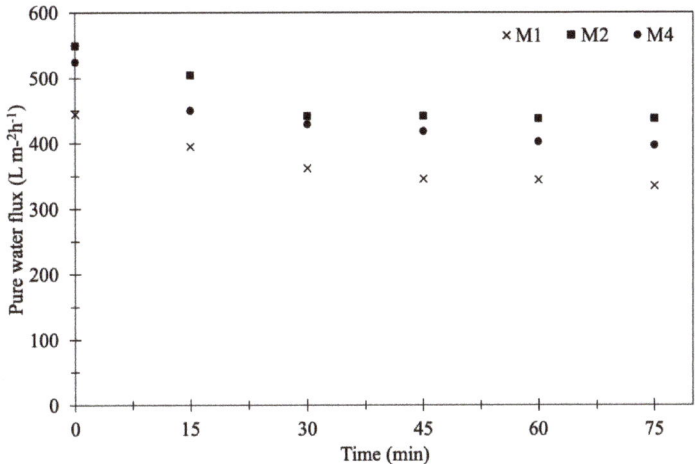

Figure 6. Water fluxes of original PVDF membrane and modified membranes (M1, M2 and M4).

In order to verify the stability and durability of the nanoparticles coated on the PVDF membrane, the original and modified PVDF membranes were tested by water filtration for the period of 75 min as illustrated in Figure 6. The stable permeate flux of M2 and M4 was higher than that of M1, which was still more than 400 L m^{-2} h^{-1}. The flux decline was observed after 30 min and then it was not significant. To determine the durability of TiO$_2$-NP and AgNP layer coated on membrane surface after 75 min water filtration, SEM-EDS of M4 was again measured together with Raman analysis and the result showed that both TiO$_2$-NP and AgNP were still presented on the membrane surface. However, the Ti and Ag concentrations before and after the pure water flux test were decreased from 4.29% to 1.01% and 3.62% to 0.55%, respectively, as shown in Table 4. It was clear that the amount of Ti and Ag might be lost with water, suggesting inadequate stability and durability of nanoparticles on the membrane due to lack of strong chemical bonding on the membrane surface. Figure 7a shows the SEM images of M4 after water filtration, which also confirmed the presence of TiO$_2$-NP and AgNP. Raman spectroscopy confirmed that TiO$_2$ was strongly deposited on modified PVDF membrane. The results of original and modified membranes exhibited a number of adsorption bands at 1703 cm^{-1}, 1650 cm^{-1}, 1453 cm^{-1}, 1100 cm^{-1}, and 840 cm^{-1}, which were attributed to C=O, CH$_2$, vibration of CH$_2$, CF$_2$, CF stretching vibration [33–35], respectively. For the modified membrane, a broad band from 600 to 400 cm^{-1} was associated with the Ti-O-Ti group and the peak of 443 cm^{-1} could be assigned to Ti-O-Ti stretching vibration [34]. No difference was observed after Raman shift at 2200 cm^{-1}. The results of Raman analysis along with SEM-EDS confirmed the deposition of TiO$_2$-NP on PVDF membrane before and after testing. To improve the durability and retention of nanoparticles on the membrane, the stabilization of those nanoparticles on membrane surface by chemical/plasma activation and crosslinked with polylactic acid might be an effective method to provide the TiO$_2$-NP and AgNP layer stability on membrane surface.

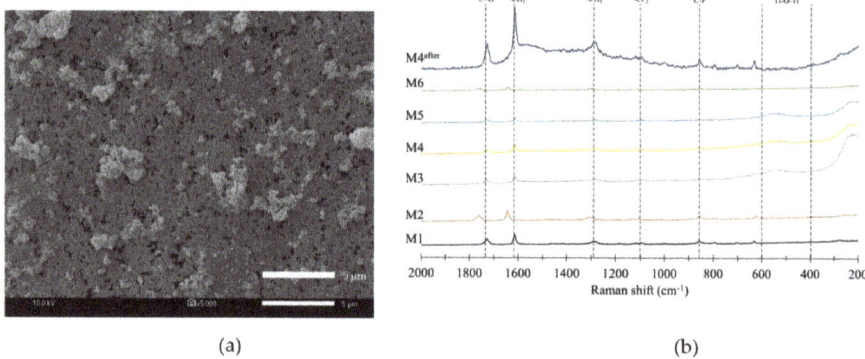

(a) (b)

Figure 7. Morphology and functional groups of membrane after pure water flux testing: (**a**) SEM image of M4; and (**b**) Raman spectra of membrane.

3.2. Antibacterial Properties of Modified Membrane

3.2.1. Antibacterial Test

(a) Effect of TiO_2-NP/AgNP concentration

Both TiO_2-NP and AgNP are well-known for their antimicrobial properties [36]. TiO_2-NP can generate reactive oxygen species (ROS), especially under ultraviolet (UV) irradiation, damaging cellular components, e.g., lipid membrane, protein, or DNA [37]. Besides oxidative stress, the released Ag^+ could inactivate several enzymes and interact with DNA, resulting in cell death [38]. In this study, the PVDF membranes coated with TiO_2-NP and AgNP were tested for their antibacterial property to reduce the number of *E. coli* cells.

The number of viable *E. coli* cells adhered onto PVDF membrane were 1.16×10^7 CFU mL^{-1}. In this study, the modified membranes showed antibacterial properties under dark condition that the modification of PVDF membrane with either 10 ppm of TiO_2-NP or AgNP significantly decreased the number of viable cells on the membrane to 0.48×10^6 and 0.89×10^6 CFU mL^{-1}, respectively (Figure 8a). In comparison with M1, the presence of both TiO_2-NP and AgNP (M4) significantly decreased the number of viable cells on the membrane by 90% (1.11×10^6 CFU mL^{-1}) (*p*-value > 0.5), comparable to those coated with only TiO_2-NP or AgNP. Increasing the amount of AgNP to 20 ppm (M5, 0.83×10^6 CFU mL^{-1}) did not significantly affect the antibacterial property of the TiO_2-NP/AgNP dip-coated PVDF membrane (M4) (*p*-value > 0.5). This is possibly due to the similar chemical composition between M4 and M5 (Table 4). On the contrary, a higher amount of TiO_2-NP (M6, 3.46×10^6 CFU mL^{-1}) substantially reduced the antibacterial activity of the TiO_2-NP/AgNP dip-coated PVDF membrane. The particular antibacterial efficiency depends on their size, shape, and surface areas that would release free radicals for the inactivation of bacterial cells [39]. Even though the higher content of Ti and Ag was detected on the M6 membrane (Table 4), too high a TiO_2-NP concentration led to the agglomeration onto PVDF membrane (Figure 3f). This could reduce the contact area between bacterial cells and nanoparticles thus decreasing the antibacterial activity. Similar results have been observed by Soo et al. (2020) in *Salmonella* Albany LCUM0022 and *Bacillus cereus* LCUM0001 under contact with Ag dopant with TiO_2 that reduces the microbial growth toward bacterial cell growth inhibition [40]. *E. coli* cell growth was also decreased with the addition of catalyst under dark condition [41]. This was the permeation of active TiO_2-NP into bacterial cell during the dark condition, resulting in the agglomeration and attachment at the bacterial membrane [42]. Of all the modified membranes, the results indicated that the mixed solution of 10 ppm TiO_2-NP and 10 ppm AgNP (M4) was suitable for the antibacterial property.

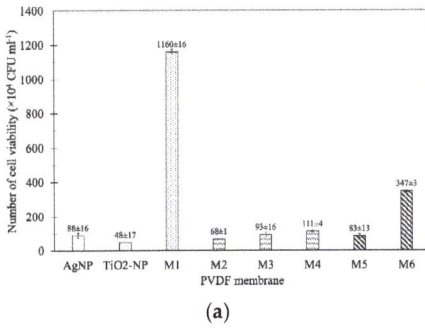 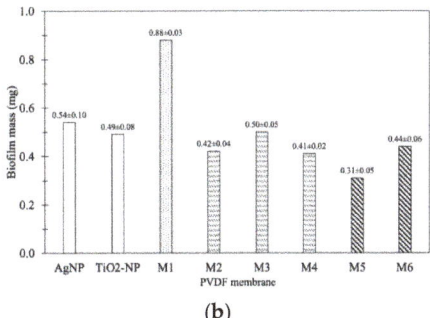

Figure 8. Antibacterial properties and biofilm inhibition test of membranes: (**a**) the number of viable *E. coli* cells and (**b**) biofilm formation on membranes.

(b) Effect of coating time.

The number of *E. coli* cells on the PVDF membranes modified with 10 ppm TiO_2-NP/10 ppm AgNP at various coating times are also presented in Figure 8b. The dipped coating of 10 ppm TiO_2-NP/10 ppm AgNP onto the PVDF membrane for 8 h (M2), 16 h (M3) and 24 h (M4), reduced the number of adhered cells by 94%, 92%, and 90%, respectively. Longer coating time resulted in the lower antibacterial efficiency of the modified membrane even though higher Ti and Ag components were observed (Table 4). This likely caused by the agglomeration of nanoparticles on the modified membrane (Figure 3); therefore, M2 with lower agglomeration could contain more nanoparticle surface area to allow more interaction sites between particles and cells, thereby showing higher antibacterial property than M3 and M4.

3.2.2. Biofilm Inhibition Test

(a) Effect of TiO_2-NP/AgNP concentration

The effect of nanoparticles-modified membrane on inhibition of biofilm formation, or antifouling property, was evaluated by using the gravimetric method. While 0.88 ± 0.03 mg of biofilm mass was detected in the unmodified PVDF membrane (M1), the coating of either TiO_2-NP or AgNP could significantly reduce the amount of biofilm to 0.49 ± 0.08 and 0.54 ± 0.10 mg, respectively (Figure 8b). The results are in agreement with the antibacterial property (Figure 8a). The coating of both nanoparticles (M4) decreased the amount of biofilm mass to 0.41 ± 0.02 mg, corresponding to 53% reduction. In comparison with M4, the increase of AgNP to 10 ppm TiO_2-NP/20 ppm AgNP (M5) further reduced the biofilm mass by 65% (0.31 ± 0.05 mg). On the other hand, the higher amount of TiO_2-NP in M6 did not improve the antifouling property (50% reduction of biomass). However, compared to either TiO_2-NP or AgNP-modified membrane, the coating of both TiO_2-NP or AgNP did not increase the antibacterial property (Figure 8a), yet the synergistic effect on the antifouling property was observed (Figure 8b). Besides the inactivation of bacterial cells, the altered surface hydrophilicity also affects the biofilm formation, i.e., antifouling property of the modified membrane. Ayyaru and Ahn (2018) [32] reported that increasing the hydrophilicity of the membrane surface could increase antifouling properties. Fouling or the formation of biofilm is initiated by the attachment of cells to the surface using the associated organelles such as flagellar and pilli [43]. Once the cells irreversibly attached to the surface, they started to produce EPS substances such as carbohydrates and proteins to form biofilm structure, which could completely block the membrane pores at this stage [44]. It has been proposed that the hydration layer on the hydrophilic surface reduces the adhesion of protein to the surface, resulting in the lower attached cells for biofilm formation [45].

(b) Effect of coating time

Since the antifouling property of PVDF membrane coated with both TiO_2-NP and AgNP (M4) was more effective than those coated with only TiO_2-NP or AgNP (Figure 8b), the effect of coating time on TiO$_2$-NP/AgNP-modified membrane was investigated. The biofilm mass of M2, M3, and M4 were 0.42 ± 0.04, 0.50 ± 0.05 and 0.41 ± 0.02 mg, respectively compared to M1 (0.8751 ± 0.03) as illustrated in Figure 8b. The PVDF membrane modified with 10 ppm TiO$_2$-NP/10 ppm AgNP reduced biofilm formation by 43–53%. Increasing the coating time (8, 16, and 24 h) did not affect the antifouling property of the modified membrane since the biofilm mass of M2, M3 and M4 were not significantly different (p >0.05). It should be noted that increasing the coating time also decreased the WCA in this study (Figure 2). Even though it has been found that lower WCA led to more hydrophilic membrane that could inhibit microorganisms adhering to the membrane [46], similar results of the modified membranes in this study suggested that more factors could play a role in antifouling properties.

4. Conclusions

In this work, the modified PVDF membrane by TiO$_2$-NP/AgNP dipped coating technique was developed to improve the hydrophilicity of the membrane along with its antibacterial and antifouling properties. The results of SEM and EDS showed that the amount of TiO$_2$-NP/AgNP on the membrane surface increased with longer coating time and higher TiO$_2$-NP/AgNP concentrations. Furthermore, increasing the coating time and TiO$_2$-NP/AgNP concentrations could decrease the WCAs of the modified membranes. Compared with the porosity of 61.33% ± 2.83% of the original PVDF membrane, the porosity of the modified membranes could be increased up to 66.74% ± 2.45% using the coating time of 24 h. At this fixed coating time, increasing the nanoparticle concentrations to 20 ppm of TiO$_2$-NP and 20 ppm of AgNP could increase the porosities to 64.10% ± 4.73% and 66.45% ± 2.78%, respectively. Noteworthy antibacterial and antifouling properties of the TiO$_2$-NP/AgNP-modified membrane were reported. The modified membranes could reduce the number of adhered *E. coli* cells by approximately 90% and the biofilm formation by at least 43%. The PVDF membrane dip-coated in 10 ppm TiO$_2$-NP and 10 ppm AgNP for 8 h showed the highest antibacterial (94% inactivation) and antifouling properties (65% reduction); the increased TiO$_2$-NP or AgNP content or coating time did not improve these properties due to the agglomeration of nanoparticles. This modified membrane is a promising alternative for improving membrane-based wastewater treatment.

Author Contributions: All authors proposed the study and participated in writing the manuscript. K.S., P.-u.S., and S.S. carried out the experiment. P.K. and P.T. discussed the antibacterial test and biofilm inhibition test. K.S., P.-u.S. and S.S. carried out the experimental design and membrane investigations. S.S. employed the AFM analysis and pure water flux experiment. S.S., P.T. and P.K. gave the final approval of the version to be submitted. S.S. is reserved for the project investigator. P.P. and M.H. gave the comments and suggestions. P.P. facilitated the membrane device system for water flux experiment. The manuscript was written, read and approved through contributions of all authors. All authors have read and agreed to the published version of the manuscript.

Funding: This work was financially supported by the Commission on Higher Education and Thailand Research Fund (MRG6280181), and Ratchadaphiseksomphot Endowment Fund, Chulalongkorn University. This research work was also partially supported by Chiang Mai University.

Acknowledgments: This work was financially supported by the Commission on Higher Education, Thailand Research Fund (MRG6280181), and Ratchadaphiseksomphot Endowment Fund, Chulalongkorn University. This research work was also partially supported by Chiang Mai University. The authors would like to thank Department of Environmental Science, Faculty of Science, Chulalongkorn University for facility provided. The authors would like to thank Prompong Pienpinijtham, Department of Chemistry, Faculty of Science, Chulalongkorn University for kindly support in Raman spectroscopy analysis.

Conflicts of Interest: The authors declare they have no competing interests.

References

1. Liu, B.; Yao, T.T.; Ren, L.X.; Zhao, Y.H.; Yuan, X.Y. Antibacterial PCL electrospun membranes containing synthetic polypeptides for biomedical purposes. *Colloids Surf. B* **2018**, *172*, 330–337. [CrossRef] [PubMed]
2. Lu, X.H.; Wang, X.K.; Guo, L.; Zhang, Q.Y.; Guo, X.M.; Li, L.S. Preparation of PU modified PVDF antifouling membrane and its hydrophilic performance. *J. Membr. Sci.* **2016**, *520*, 933–940. [CrossRef]

3. Chew, N.G.P.; Zhao, S.S.; Loh, C.H.; Permogorov, N.; Wang, R. Surfactant effects on water recovery from produced water via direct-contact membrane distillation. *J. Membr. Sci.* **2017**, *528*, 126–134. [CrossRef]
4. Miao, R.; Wang, L.; Feng, L.; Liu, Z.W.; Lv, Y.T. Understanding PVDF ultrafiltration membrane fouling behaviour through model solutions and secondary wastewater effluent. *Desalin. Water Treat.* **2014**, *52*, 5061–5067. [CrossRef]
5. Zeng, K.L.; Zhou, J.; Cui, Z.L.; Zhou, Y.; Shi, C.; Wang, X.Z.; Zhou, L.Y.; Ding, X.B.; Wang, Z.H.; Drioli, E. Insight into fouling behavior of poly(vinylidene fluoride) (PVDF) hollow fiber membranes caused by dextran with different pore size distributions. *Chin. J. Chem. Eng.* **2018**, *26*, 268–277. [CrossRef]
6. Chew, N.G.P.; Zhao, S.S.; Wang, R. Recent advances in membrane development for treating surfactant- and oil-containing feed streams via membrane distillation. *Adv. Colloid Interface Sci.* **2019**, *273*, 102022. [CrossRef]
7. Zuo, G.Z.; Wang, R. Novel membrane surface modification to enhance anti-oil fouling property for membrane distillation application. *J. Membr. Sci.* **2013**, *447*, 26–35. [CrossRef]
8. Aslam, M.; Ahmad, R.; Kim, J. Recent developments in biofouling control in membrane bioreactors for domestic wastewater treatment. *Sep. Purif. Technol.* **2018**, *206*, 297–315. [CrossRef]
9. Wu, L.G.; Zhang, X.Y.; Wang, T.; Du, C.H.; Yang, C.H. Enhanced performance of polyvinylidene fluoride ultrafiltration membranes by incorporating TiO$_2$/graphene oxide. *Chem. Eng. Res. Des.* **2019**, *141*, 492–501. [CrossRef]
10. Alammar, A.; Park, S.H.; Williams, C.J.; Derby, B.; Szekely, G. Oil-in-water separation with graphene-based nanocomposite membranes for produced water treatment. *J. Membr. Sci.* **2020**, *603*, 118007. [CrossRef]
11. Wang, F.; Dai, J.W.; Huang, L.Q.; Si, Y.; Yu, J.Y.; Ding, B. Biomimetic and Superelastic Silica Nanofibrous Aerogels with Rechargeable Bactericidal Function for Antifouling Water Disinfection. *ACS Nano* **2020**, *14*, 8975–8984. [CrossRef] [PubMed]
12. Penboon, L.; Khrueakham, A.; Sairiam, S. TiO$_2$ coated on PVDF membrane for dye wastewater treatment by a photocatalytic membrane. *Water Sci. Technol.* **2019**, *79*, 958–966. [CrossRef] [PubMed]
13. Pi, J.K.; Yang, H.C.; Wan, L.S.; Wu, J.; Xu, Z.K. Polypropylene microfiltration membranes modified with TiO$_2$ nanoparticles for surface wettability and antifouling property. *J. Membr. Sci.* **2016**, *500*, 8–15. [CrossRef]
14. Madaeni, S.S.; Ghaemi, N. Characterization of self-cleaning RO membranes coated with TiO$_2$ particles under UV irradiation. *J. Membr. Sci.* **2007**, *303*, 221–233. [CrossRef]
15. Moriyama, A.; Yamada, I.; Takahashi, J.; Iwahashi, H. Oxidative stress caused by TiO$_2$ nanoparticles under UV irradiation is due to UV irradiation not through nanoparticles. *Chem. Biol. Interact.* **2018**, *294*, 144–150. [CrossRef]
16. Chew, N.G.P.; Zhang, Y.J.; Goh, K.; Ho, J.S.; Xu, R.; Wang, R. Hierarchically Structured Janus Membrane Surfaces for Enhanced Membrane Distillation Performance. *ACS Appl. Mater. Interfaces* **2019**, *11*, 25524–25534. [CrossRef]
17. Huang, L.C.; Zhao, S.; Wang, Z.; Wu, J.H.; Wang, J.X.; Wang, S.C. In situ immobilization of silver nanoparticles for improving permeability, antifouling and anti-bacterial properties of ultrafiltration membrane. *J. Membr. Sci.* **2016**, *499*, 269–281. [CrossRef]
18. Qi, L.B.; Liu, Z.Y.; Wang, N.; Hu, Y.X. Facile and efficient in situ synthesis of silver nanoparticles on diverse filtration membrane surfaces for antimicrobial performance. *Appl. Surf. Sci.* **2018**, *456*, 95–103. [CrossRef]
19. Wang, L.; Ali, J.; Zhang, C.B.; Mailhot, G.; Pan, G. Simultaneously enhanced photocatalytic and antibacterial activities of TiO$_2$/Ag composite nanofibers for wastewater purification. *J. Environ. Chem. Eng.* **2020**, *8*, 102104. [CrossRef]
20. Zou, Y.J.; Liang, J.; She, Z.; Kraatz, H.B. Gold nanoparticles-based multifunctional nanoconjugates for highly sensitive and enzyme-free detection of *E.coli* K12. *Talanta* **2019**, *193*, 15–22. [CrossRef]
21. Behboudi, A.; Jafarzadeh, Y.; Yegani, R. Enhancement of antifouling and antibacterial properties of PVC hollow fiber ultrafiltration membranes using pristine and modified silver nanoparticles. *J. Environ. Chem. Eng.* **2018**, *6*, 1764–1773. [CrossRef]
22. Rompre, A.; Servais, P.; Baudart, J.; de Roubin, M.R.; Laurent, P. Detection and enumeration of coliforms in drinking water: Current methods and emerging approaches. *J. Microbiol. Methods* **2002**, *49*, 31–54. [CrossRef]
23. Huang, J.; Wang, H.T.; Zhang, K.S. Modification of PES membrane with Ag-SiO$_2$: Reduction of biofouling and improvement of filtration performance. *Desalination* **2014**, *336*, 8–17. [CrossRef]

24. Ghaemi, N. Novel antifouling nano-enhanced thin-film composite membrane containing cross-linkable acrylate-alumoxane nanoparticles for water softening. *J. Colloid Interface Sci.* **2017**, *485*, 81–90. [CrossRef] [PubMed]
25. Maheswari, P.; Prasannadevi, D.; Mohan, D. Preparation and performance of silver nanoparticle incorporated polyetherethersulfone nanofiltration membranes. *High Perform. Polym.* **2013**, *25*, 174–187. [CrossRef]
26. Gebru, K.A.; Das, C. Removal of bovine serum albumin from wastewater using fouling resistant ultrafiltration membranes based on the blends of cellulose acetate, and PVP-TiO$_2$ nanoparticles. *J. Environ. Manag.* **2017**, *200*, 283–294. [CrossRef] [PubMed]
27. Yang, Z.; Peng, H.D.; Wang, W.Z.; Liu, T.X. Crystallization behavior of poly(epsilon-caprolactone)/layered double hydroxide nanocomposites. *J. Appl. Polym. Sci.* **2010**, *116*, 2658–2667.
28. Chew, N.G.P.; Zhao, S.S.; Malde, C.; Wang, R. Superoleophobic surface modification for robust membrane distillation performance. *J. Membr. Sci.* **2017**, *541*, 162–173. [CrossRef]
29. Chew, N.G.P.; Zhao, S.S.; Malde, C.; Wang, R. Polyvinylidene fluoride membrane modification via oxidant-induced dopamine polymerization for sustainable direct-contact membrane distillation. *J. Membr. Sci.* **2018**, *563*, 31–42. [CrossRef]
30. Abedini, R.; Mousavi, S.M.; Aminzadeh, R. A novel cellulose acetate (CA) membrane using TiO$_2$ nanoparticles: Preparation, characterization and permeation study. *Desalination* **2011**, *277*, 40–45. [CrossRef]
31. Rahimpour, A.; Jahanshahi, M.; Rajaeian, B.; Rahimnejad, M. TiO$_2$ entrapped nano-composite PVDF/SPES membranes: Preparation, characterization, antifouling and antibacterial properties. *Desalination* **2011**, *278*, 343–353. [CrossRef]
32. Ayyaru, S.; Ahn, Y.H. Fabrication and separation performance of polyethersulfone/sulfonated TiO$_2$ (PES-STiO$_2$) ultrafiltration membranes for fouling mitigation. *J. Ind. Eng. Chem.* **2018**, *67*, 199–209. [CrossRef]
33. Lou, L.H.; Kendall, R.J.; Smith, E.; Ramkumar, S.S. Functional PVDF/rGO/TiO$_2$ nanofiber webs for the removal of oil from water. *Polymer* **2020**, *186*, 122028. [CrossRef]
34. Lihua, L.; Kendall, R.J.; Ramkumar, S. Comparision of hydrophilic PVA/TiO$_2$ and hydrophobic PVDF/TiO$_2$ microfiber webs on the dye pollutant photo-catalyzation. *J. Environ. Chem. Eng.* **2020**, *9*, 103914.
35. Chen, X.J.; Huan, G.; An, C.J.; Feng, R.F.; Wu, Y.H.; Huang, C. Plasma-induced PAA-ZnO coated PVDF membrane for oily wastewater treatment: Preparation, optimization, and characterization through Taguchi OA design and synchrotron-based X-ray analysis. *J. Membr. Sci.* **2019**, *582*, 70–82. [CrossRef]
36. Bonan, R.F.; Mota, M.F.; Farias, R.M.D.; da Silva, S.D.; Bonan, P.R.F.; Diesel, L.; Menezes, R.R.; Perez, D.E.D. In vitro antimicrobial and anticancer properties of TiO$_2$ blow-spun nanofibers containing silver nanoparticles. *J. Ind. Eng. Chem.* **2019**, *104*, 109876. [CrossRef]
37. Sunada, K.; Kikuchi, Y.; Hashimoto, K.; Fujishima, A. Bactericidal and detoxification effects of TiO$_2$ thin film photocatalysts. *Environ. Sci. Technol.* **1998**, *32*, 726–728. [CrossRef]
38. Prabhu, S.; Poulose, E.K. Silver nanoparticles: Mechanism of antimicrobial action, synthesis, medical applications, and toxicity effects. *Int. Nano Lett.* **2012**, *2*, 32. [CrossRef]
39. Takeshima, T.; Tada, Y.; Sakaguchi, N.; Watari, F.; Fugetsu, B. DNA/Ag nanoparticles as antibacterial agents against gram-negative bacteria. *Nanomaterials* **2015**, *5*, 284–297. [CrossRef]
40. Soo, J.Z.; Chai, L.C.; Ang, B.C.; Ong, B.H. Enhancing the Antibacterial Performance of Titanium Dioxide Nanofibers by Coating with Silver Nanoparticles. *ACS Appl. Nano Mater.* **2020**, *3*, 5743–5751. [CrossRef]
41. Yang, L.; Ye, F.Y.; Liu, P.; Wang, F.Z. The Visible-Light Photocatalytic Activity and Antibacterial Performance of Ag/AgBr/TiO$_2$ Immobilized on Activated Carbon. *Photochem. Photobiol.* **2016**, *92*, 800–807. [CrossRef] [PubMed]
42. Dalai, S.; Pakrashi, S.; Kumar, R.S.S.; Chandrasekaran, N.; Mukherjee, A. A comparative cytotoxicity study of TiO$_2$ nanoparticles under light and dark conditions at low exposure concentrations. *Toxicol. Res.* **2012**, *1*, 116–130. [CrossRef]
43. Reisner, A.; Haagensen, J.A.J.; Schembri, M.A.; Zechner, E.L.; Molin, S. Development and maturation of *Escherichia coli* K-12 biofilms. *Mol. Microbiol.* **2003**, *48*, 933–946. [CrossRef] [PubMed]
44. Malamis, S.; Andreadakis, A. Fractionation of proteins and carbohydrates of extracellular polymeric substances in a membrane bioreactor system. *Bioresour. Technol.* **2009**, *100*, 3350–3357. [CrossRef]

45. Jimenez-Pardo, I.; van der Ven, L.G.J.; van Benthem, R.A.T.M.; de With, G.; Esteves, A.C.C. Hydrophilic self-replenishing coatings with long-term water stability for anti-fouling applications. *Coatings* **2018**, *8*, 184. [CrossRef]
46. Zhu, J.Y.; Hou, J.W.; Zhang, Y.T.; Tian, M.M.; He, T.; Liu, J.D.; Chen, V. Polymeric antimicrobial membranes enabled by nanomaterials for water treatment. *J. Membr. Sci.* **2018**, *550*, 173–197. [CrossRef]

Publisher's Note: MDPI stays neutral with regard to jurisdictional claims in published maps and institutional affiliations.

© 2020 by the authors. Licensee MDPI, Basel, Switzerland. This article is an open access article distributed under the terms and conditions of the Creative Commons Attribution (CC BY) license (http://creativecommons.org/licenses/by/4.0/).

Review

A Review on the Mechanism, Impacts and Control Methods of Membrane Fouling in MBR System

Xianjun Du [1,2,3,4,*], Yaoke Shi [1], Veeriah Jegatheesan [2] and Izaz Ul Haq [1]

1. College of Electrical and Information Engineering, Lanzhou University of Technology, Lanzhou 730050, China; yaoke_shi@163.com (Y.S.); Izaz.lut@gmail.com (I.U.H.)
2. School of Engineering, RMIT University, Melbourne 3000, Australia; jega.jegatheesan@rmit.edu.au
3. Key Laboratory of Gansu Advanced Control for Industrial Processes, Lanzhou University of Technology, Lanzhou 730050, China
4. National Demonstration Center for Experimental Electrical and Control Engineering Education, Lanzhou University of Technology, Lanzhou 730050, China
* Correspondence: 27dxj@163.com

Received: 28 December 2019; Accepted: 30 January 2020; Published: 4 February 2020

Abstract: Compared with the traditional activated sludge process, a membrane bioreactor (MBR) has many advantages, such as good effluent quality, small floor space, low residual sludge yield and easy automatic control. It has a promising prospect in wastewater treatment and reuse. However, membrane fouling is the biggest obstacle to the wide application of MBR. This paper aims at summarizing the new research progress of membrane fouling mechanism, control, prediction and detection in the MBR systems. Classification, mechanism, influencing factors and control of membrane fouling, membrane life prediction and online monitoring of membrane fouling are discussed. The research trends of relevant research areas in MBR membrane fouling are prospected.

Keywords: membrane fouling; influencing factors; control method

1. Introduction

Membrane bioreactors (MBRs) combine both biological treatment and physical separation (using membranes) of various pollutants to treat domestic and industrial liquid wastes. Due to the combination of the above-mentioned processes, MBRs produce treated effluents of higher quality compared to the conventional activated sludge process [1]. MBRs are simple to operate when experienced operators are employed and produce less sludge. They also have a very small footprint which is very valuable when MBRs are installed in dense urban areas where space is at a premium. The advantages mentioned above, along with the ever-decreasing cost of membrane materials and the increasingly stringent requirements of treated effluent quality, mean that MBR technology is more and more widely applied in wastewater treatment [2]. However, membrane fouling affects the operating flux and the life of membranes. There is no unified statement about the mechanisms of membrane fouling, but from the analysis of the causes of membrane fouling, the following mechanisms of membrane fouling in MBR have been proposed and verified: (i) narrowing of membrane pores; (ii) the adsorption of the solute in the solution by the membrane [3]; (iii) the deposition of the (activated) sludge floc on the membrane surface [4]; and (iv) the compaction of the filter cake layer on the membrane surface. These mechanisms alone or together play a leading role at different stages of the membrane filtration process. As various factors govern the operating cost of a membrane system, such as power requirements, costs of power, labor, materials, membrane cleaning, scale inhibition and membrane life and replacement, some limitations remain in using membranes for water and wastewater treatment [5]. Once membrane fouling occurs, it will reduce permeate flux, increase feed pressure, reduce productivity, increase system downtime, increase membrane maintenance and operation costs due to membrane cleaning, and decrease the lifespan of the membrane modules [6]. Thus, the main challenge in the application of MBRs is to find a solution to the fouling of membranes [7,8]. Most of the existing papers on membrane fouling review a specific kind of research direction in a specific area (see Appendix A), which is not comprehensive enough. However, membrane fouling itself involves a wide range of content and covers a lot of knowledge basics. Those who want to promote the research on membrane fouling must fully understand membrane fouling. In this paper, the traditional research content of membrane fouling and the latest research results are integrated together, including every aspect of membrane fouling research by many scholars and researchers in the past decades, which implements the most complete discussion of influencing factors, mechanisms and control methods of membrane fouling by far. It provides a more comprehensive and systematic reference for follow-up studies on membrane fouling and relevant areas.

2. Classification of Membrane Fouling

Membrane fouling can be classified into internal fouling, external fouling and concentration polarization fouling. The fouling caused by the deposition as well the adsorption of solutes and colloidal particles on the interior of the membrane pores is called internal fouling and sometimes referred to as pore blocking [9]. The deposition of particles, colloids and macromolecules on the membrane surface is called external fouling. External fouling forms a fouling layer on the membrane surface. The fouling layer can be classified as gel layer or cake layer. The gel layer is formed by the deposition of macromolecules, colloids and inorganic solutes on the surface of the membrane due to the pressure difference between the feed and permeate sides of the membrane. The cake layer is formed by the accumulation of solids on the membrane surface [10]. Concentration polarization refers to the accumulation of solutes and ions in the thin liquid layer adjacent to the membrane surface [11], which is an inherent phenomenon in the membrane filtration process. Concentration polarization increases the flow resistance and decreases the membrane flux. The concentration polarization layer is determined by the convective shear force [12]. Increasing the convective velocity can alleviate membrane resistance caused by concentration polarization.

Membrane fouling has traditionally been divided into reversible fouling and irreversible fouling according to the degree of removal of foulants [13]. Reversible fouling refers to the part of the foulants that can be removed by physical means such as backwashing or intermittent operation of membranes under cross-flow filtration. Non-reversible fouling refers to the fouling that needs chemical cleaning and cannot be removed by physical cleaning [14]. It is generally believed that reversible fouling is caused by loose deposition of contaminants on the surface of the membrane and irreversible fouling is caused by the blockage of membrane pores and strong adhesion of contaminants to the surface of the membrane. Many studies have pointed out that the formation and compaction of the cake layer is the main form of membrane fouling compared to membrane pore blocking [15]. Table 1 shows the onset of various fouling of membranes in an MBR.

Table 1. Onset of various fouling of membrane in a membrane bioreactor (MBR).

Fouling Type	Rate of Fouling (Pa.min^{-1})	Onset of Fouling
Reversible fouling	10–100	10 min
Irreversible fouling (removed by maintenance chemical cleaning)	1–10	1–2 weeks
Irreversible fouling (removed by mandatory chemical cleaning)	0.1–1	6–12 months
Non-restorable fouling	0.01–0.1	A few years

According to the composition of pollutants, membrane fouling can be divided into organic fouling, inorganic fouling and biofouling. Organic fouling is caused by organic macromolecules. Wang et al. [16] found that organic macromolecular polymer clusters (BPCs) are important contaminants. Analysis shows that BPCs are less than 50 μm in diameter, which are significantly different from activated sludge floc particles. Lin et al. [17] found that the organic matter content in the activated sludge supernatant was significantly higher than that in the MBR effluent, and the high content of organic matter in the supernatant was considered to contain BPCs. BPCs act like glue, which helps the sludge to adhere to the surface of the membrane and form a cake layer. Wang et al. [18] studied the formation process and fouling characteristics of dynamic membranes as well as an improved self-forming dynamic membrane bioreactor (SF-DMBR) for the recovery of organic matter in wastewater and to evaluate its properties. The results showed that 80% of the organic matter in the wastewater can be recovered faster in the case of continuous operation. Inorganic foulants include struvite, $K_2NH_4PO_4$ and $CaCO_3$. Biofouling is caused by the interaction between biological substances and membranes. Gao et al. [19] found that about 65% of the particles in the membrane cake layer are smaller than the pore size of the membrane (0.1–0.4 μm), so that they can pass through the membrane pores and can block the membrane pores. Biofouling also includes adsorption of extracellular polymers (EPS) and microbial metabolites (SMP), which are produced by microbial secretion, onto the membrane surface and membrane pores [20]. The microbial colony structures in the membrane cake layer and in the mixed liquor are significantly different. Some strains will preferentially adsorb onto the membrane surface due to the secretion of more EPS, resulting in serious biofouling [21].

3. Factors Affecting the Fouling of Membranes

There are many factors that cause membrane fouling, including the material of the membrane module, the pressure difference across the membrane during filtration, the cross-flow velocity, the hydraulic retention time (HRT), the sludge retention time (SRT), microbial polymerization and dissolution processes, and mixing [22]. These factors alone or in combination provide conditions for membrane fouling, or contribute directly or indirectly to membrane fouling. Understanding and mastering the effects of various factors on membrane fouling is essential to prevent, control and predict membrane fouling [23]. The influencing factors on membrane fouling are shown in Table 2.

3.1. Influence of Membrane Intrinsic Properties on Membrane Fouling

Intrinsic properties of the membrane that affect membrane fouling include material, hydrophilicity/hydrophobicity, surface charge, roughness, pore size, porosity, and structure of the membrane module.

3.1.1. Effect of Membrane Material on Membrane Fouling

Commonly used membranes are mainly classified into organic membranes, ceramic membranes, and metal membranes according to the type of materials used to synthesize membranes. Among them, the organic membranes have a low cost and mature manufacturing process, and are currently the most widely used [24]. However, they have low strength as well as a short life and are easily fouled. Common organic membranes include polyethylene (PE), polysulfone (PS), polyethersulfone (PES), polyacrylonitrile (PAN) and polyvinylidene fluoride (PVDF). Among them, PVDF membrane has a higher anti-fouling ability. Compared with organic membranes, ceramic membranes and metal membranes have (i) better mechanical properties, (ii) resistance to high temperatures, and (iii) high flux; however, these two materials are difficult to manufacture and are expensive [25]. In the case of the same operating conditions [26], the PAN membrane pollutes slower than the PES membrane. Researchers [27] found that metal membranes are easier to recover from fouling than organic membranes.

Table 2. Factors affecting membrane fouling.

Factor	Influence	Type of Wastewater
Membrane structure properties	The formation of the cake layer can be observed in the organic fouling, and inorganic fouling did not easily cause membrane fouling.	-
Material characteristics	The protein in the EPS was more than the polysaccharide, and the viscosity of the liquid increased.	Hot white pulp wastewater
	Increased SMP, increased filtration resistance, and deterioration of membrane due to fouling.	Domestic wastewater
	Supernatant SMP had more protein than polysaccharides, the viscosity increased, and the cake layer was easy to form.	Industrial waste
Operating condition	When SRT increased, SMP and sludge viscosity increased.	Low concentration wastewater
	At 30 and 50 d, the activated sludge floc increased, the low fouling rate SRT was too small, the SMP increased, and the fouling accelerated.	Municipal wastewater
	If it was too large, MLSS, SMP and other microbial products increased.	-
	HRT declined, protein substances in SMP increased, and EPS concentration increased.	Low concentration wastewater
	HRT decreased, filtration resistance increased, and granular sludge particle size decreased.	Artificial wastewater
	Small flocs increased under high temperature conditions, SMP, EPS increase, filter cake layer was easy to form	Evaporator condensate
	When the temperature went up, the membrane fouling resistance increased, and the protein content in EPS increased.	Hot pulping press

3.1.2. Effect of Hydrophilicity/Hydrophobicity on Membrane Fouling

With the continuous application of membrane bioreactors, membrane fouling has become a major bottleneck limiting its further development [28,29]. At the beginning of operation, hydrophilic organic matter will be the dominant pollutant; however, the interaction force between hydrophobic organic matter and the membrane is significantly greater than the interaction force between hydrophilic organic matter and the membrane, resulting in hydrophobic organic matter becoming a dominant pollutant in the later stage of operation [30]. Among them, hydrophilic carbohydrate organic matter and hydrophobic humic organic matter abundantly present in the wastewater to be treated are key substances causing membrane fouling [31]. Therefore, it is important to find out how the pro-/hydrophobic organic matter contaminates the membrane to prevent and control the membrane fouling.

The hydrophilicity/hydrophobicity of the membrane is usually characterized by the contact angle θ. The larger the value of θ, the stronger the hydrophobicity of the membrane surface. The value of θ angle has a certain relationship with the morphology of the membrane surface and the pore size of the membrane. The hydrophilicity/hydrophobicity of the membrane material has a great influence on the anti-fouling performance of the membrane [32]. The hydrophilic membrane is less affected by adsorption, has a larger membrane flux, and has superior anti-fouling properties compared to the hydrophobic membrane. However, some researchers have concluded that the most hydrophilic PES membrane suffers from the most serious membrane fouling, which may be related to the maximum membrane pore opening of PES. It is worth noting that the hydrophilicity/hydrophobicity of the membrane usually only has a significant effect on membrane fouling at the initial stage of filtration. After the initial fouling, the chemical properties of the foulants will replace the chemical properties of the membrane itself as the main influencing factor [33].

The natural organic matter present in wastewater is divided into strongly hydrophobic, weakly hydrophobic, polar hydrophilic and neutral hydrophilic organic matter, and is filtered separately. It is found that the most important organic matter causing the decrease of membrane flux is neutral hydrophilic organic matter. It is believed that the presence of more hydrophilic organic matter in the raw water causes more serious membrane fouling [19]. However, some researchers have reached the opposite conclusion [15,34]. Their experiments showed that the hydrophobic organic matter is the main factor causing the decline in flux.

3.1.3. Effect of Membrane Surface Charge on Membrane Fouling

When the membrane surface charge is the same as the charge of the pollutants present in the wastewater, it can improve the membrane surface contamination and increase the membrane flux. In general, the colloidal particles in the aqueous solution are negatively charged. So, if the material having a negative potential is used as the membrane material, it can prevent membrane fouling due to the repellent effect of similar charges.

Lin et al. [35] investigated the effects of operating pressure difference, cross-flow rate, feed solution concentration and operating temperature on the flow potential of suspended-growth (SG) nanofiltration membranes; the membranes were filtering NaCl, Na_2CO_3 and $CuCl_2$ solutions and they found that SG membranes have negative surface charges. At the same time, the absolute values of electric density, zeta potential and charge density increased with the increase of cation or anion valence state. The higher the operating pressure difference and the concentration of the liquid solution, the lower the velocity of the cross flow and the influence of the valence state of the ions. The cation of the same valence state have a greater effect on the charge performance of the membrane than the anions [36]. Figure 1 shows the molecular weight distribution range of the simulated hydrophilic (HPI) and hydrophobic (HPO) organics in the secondary treated effluent.

As can be seen from the figure, the actual molecular weight distribution range of hydrophilic/hydrophobic organics is relatively small, mainly concentrated in the range of <10,000 Dalton. The proportion of organic molecules with HPO < 10,000 Dalton is close to 70%, and especially the proportion of organic molecules with HPI < 10,000 is up to 80%.

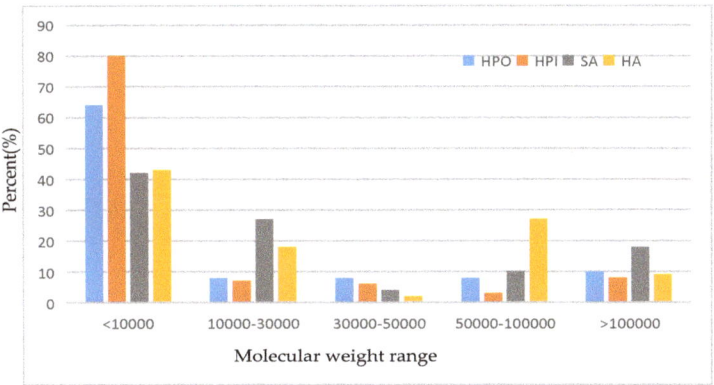

Figure 1. Molecular weight distribution of different organic matters in secondary treated effluent. (Note: HPO-hydrophobic; HPI-hydrophilic; SA-sodium alginate; HA-humic acid).

3.1.4. Effect of Membrane Pore Size, Distribution and Structure on Membrane Fouling

In a treatment study of wastewater containing micropollutants [37], it was found that the pore size of the membrane is the main factor affecting the membrane flux and the turbidity removal rate of the wastewater. The larger the membrane pore size, the more serious the fouling of the membrane, and faster the flux decay as well as the lower the removal rate. Membrane pore size is an important indicator that directly affects the separation performance of the membrane. In the membrane filtration process, membrane surface or membrane pores easily become sites for adsorption and deposition; they are also blocked by tiny particles or solute macromolecules present in water [38]. According to the characteristics of water sources with micropollutants, the membrane fouling caused by small molecular substances cannot be ignored. When using MBR to treat water with micropollutants, it is important to select membranes with appropriate pore size for process operation and membrane fouling control. The change of transmembrane pressure difference and water flux with time in three kinds of membranes with different pore sizes is shown in Figure 2. As the pore size of the membrane increases, the transmembrane pressure difference of the membrane increases rapidly [39].

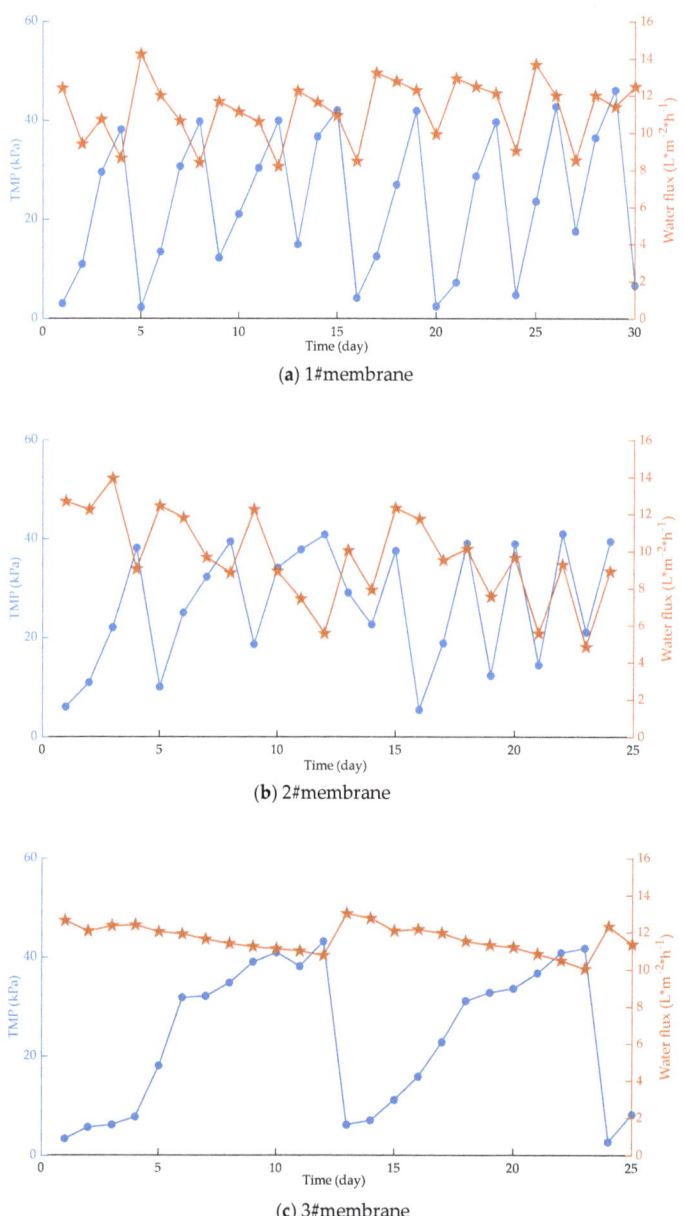

(a) 1#membrane

(b) 2#membrane

(c) 3#membrane

Figure 2. Temporal variations of flux and transmembrane pressure (TMP) when membranes with different pore sizes are used. (**a**) 1#membrane; (**b**) 2#membrane; (**c**) 3#membrane.

It is found from Figure 1 that the degree of membrane fouling is aggravated with the extension of the running time, resulting in an increase in the transmembrane pressure difference and a decrease in the water flux. In addition, the smaller the membrane pore size, the slower the increase in transmembrane pressure difference, the longer the membrane cleaning cycle time, the larger the membrane pore size, the more serious the membrane fouling, and the shorter the membrane cleaning cycle time. However, due to high membrane surface porosity and fiber-interwoven network-like pore structure, (i) the membrane can be cleaned effectively, (ii) the performance of the membrane can be recovered very well after repeated cleaning, and (iii) the water flux of the membrane changes minimally [40].

3.1.5. Effect of Porosity and Roughness on Membrane Fouling

Membrane porosity and roughness also have a potential impact on membrane fouling behavior. Generally, the larger the porosity, the smaller the transmembrane pressure (TMP). However, as the porosity changes, the surface properties of the membrane, such as roughness, also change. This in turn changes the possibility of adsorbing contaminants on the membrane surface. The organic membrane porosity is usually higher than that of the inorganic membrane, but the flux is often lower than that of the inorganic membrane [41,42]. When the membrane surface roughness is large, the membrane is more susceptible to fouling [43].

3.1.6. Effect of Membrane Module Structure on Membrane Fouling

The membrane module is the core of membrane separation technology. For a membrane separation process, not only membranes with excellent separation characteristics but also membrane modules and devices with compact structure and stable performance must be applied in large-scale industrial processes [44]. In the process of studying PVC membrane materials [45], it was found that under the same operating conditions, the membrane bioreactor with a vertical membrane module has a better flow state, and the shearing effect of the gas and liquid two-phase flow generated by aeration and scouring is much stronger than that of a horizontal device. In addition, the bottom of the horizontal device is perforated and aerated, the PVC film is thicker, and the sludge deposition on the side of the aeration hole is also serious, so the membrane fouling rate is faster and the corresponding operation period is shorter [46]. Under the same operating conditions, the impact of different placement modes on membrane fouling of two typical systems is shown in Figure 3.

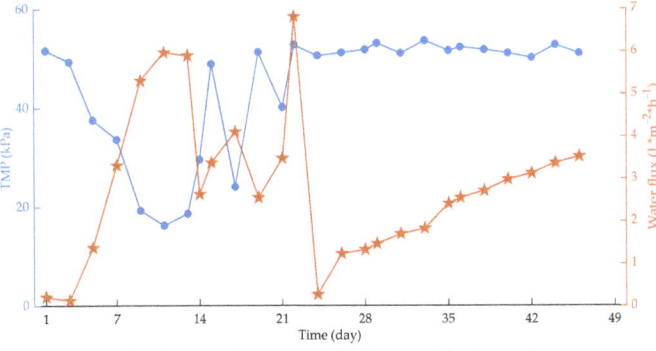

(a) The membrane assembly is placed horizontally

Figure 3. *Cont.*

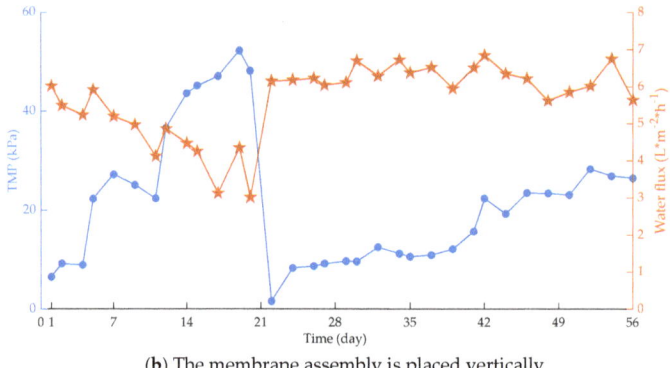

(**b**) The membrane assembly is placed vertically

Figure 3. Influence of membrane placement on membrane fouling. (**a**) The membrane assembly is placed horizontally; (**b**) The membrane assembly is placed vertically.

It can be seen that under the same operating conditions, the flow condition of the membrane bioreactor with the membrane component placed vertically is better, and the shear effect of gas–liquid two-phase flow generated by aeration scour is much stronger than that of the horizontal release device.

Since FMX maintains high flux and recovery rate under the most challenging conditions, it has been used for wastewater treatment, separation and dewatering in manufacturing processes and recovery applications in various industries. Chen et al. [47] compared the membrane filtration process using FMX rotating disc plate, hollow fiber, tubular and filter cup membrane modules; the results showed that the FMX membrane module yielded higher membrane flux with lower fouling of the membrane, followed by the hollow fiber membrane, tubular, and the filter cup membrane modules when operated under dead-end filtration mode. Results are shown in Figure 4.

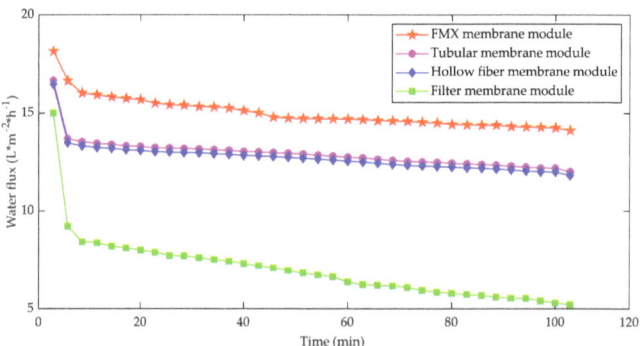

Figure 4. Variation of permeate flux of four membrane modules.

3.2. Effect of Operating Conditions on Membrane Fouling

Operating conditions affecting membrane fouling include membrane flux, TMP, aeration, cross-flow velocity (CFV), SRT, HRT, and temperature. Furthermore, mode of operation, including influent water quality and sludge loading, directly affects membrane fouling factors.

3.2.1. Effect of Membrane Flux and TMP on Membrane Fouling

In membrane filtration operations, membrane flux and TMP are two quantities associated with each other [48]. If other conditions remain the same, in order to obtain higher membrane flux,

TMP must be increased. Conversely, if TMP is increased or decreased, membrane flux will change accordingly [49]. MBR has two modes of operation: constant flux and constant pressure. Many studies have demonstrated the existence of critical fluxes; MBRs are operated below critical to avoid excessive fouling of the membrane during the initial phase of operation [50]. Under low-pressure operation, the initial filter cake layer formed is thin or only has a reversible concentration polarization layer, so the membrane fouling is not significant. However, operating the MBR above the critical TMP makes the initial filter cake layer thicker or forms a concentrated polarization layer. It is also converted into a dense filter cake layer [51], which increases membrane fouling. Lowering the initial TMP can reduce membrane fouling and slow down the rate of membrane flux decline [52]. In addition, membrane fouling is relatively slow in constant flux operation compared to constant TMP operation, but membrane flux recovery is poor after cleaning. This may be due to the continuous densification of the fouling layer on the membrane surface during constant flux operation [53].

3.2.2. Effect of Aeration and CFV on Membrane Fouling

Aeration is an important parameter in the operation of MBR. It not only provides the oxygen necessary for metabolism of activated sludge, but also washes the surface of the membrane, avoids the deposition of pollutants and slows the fouling of the membrane. Therefore, membrane bioreactors often use relatively larger aeration [54]. The change of aeration volume in MBR will cause changes in the characteristics of effluent water quality and sludge mixture. When the aeration rate increases, the effluent quality will improve and the removal rates of COD (chemical oxygen demand) and NH_4^+-N will increase, but the sludge floc size will decrease, the impact on sludge concentration and sludge load is weak. As the amount of aeration increases, the total amount and composition of SMP will change [55]. Among them, the protein/polysaccharide value has an important influence on the sludge properties, which in turn affect the physical, chemical and biological properties of the sludge mixture and ultimately affect the fouling rate. The occurrence of membrane fouling will lead to a decrease in membrane flux to some extent [56]. Moderate aeration can reduce membrane fouling and increase membrane flux to some extent [57]. Zhang et al. [58] found that when the aeration intensity increased to a certain extent, membrane pore adsorption, clogging and membrane gel layer resistance became the main membrane resistance and the fouling rate increased. Therefore, there is an optimum aeration intensity for the operation of MBR.

The change in CFV changes the diffusion caused by shear to affect the migration of particles from the surface of the membrane, which in turn affects the thickness of the cake layer. The membrane flux increases approximately linearly with increasing CFV. However, the CFV is not as a large factor as the aeration, and the better the membrane filtration performance will be after CFV exceeds the critical value. However, when the CFV exceeds a certain threshold, the TMP will increase as the CFV increases [59]. This is because higher CFV reduces the deposits of larger particles and allows the filter cake layer to consist primarily of small particles. Those particles are more compact and lead to higher TMP. Moreover, too large a CFV will cause the sludge particles to break, which will make the filter cake layer more dense, and also stimulate the release of EPS and increase membrane fouling [60].

3.2.3. Effect of SRT and HRT on Membrane Fouling

SRT can affect MLSS (mixed liquid suspended solids), sludge composition, EPS and other parameters which are important operating conditions affecting membrane fouling rate in an MBR. HRT has an indirect effect on membrane fouling. First, changes in HRT will directly lead to changes in membrane flux, which in turn will change the state of membrane filtration and affect the rate of membrane fouling [61].

When investigating the effect of HRT on membrane fouling in a split-type anaerobic membrane bioreactor (AnMBR) treating beer wastewater, it was found that shorter HRT can produce higher OLR and F/M, which in turn affects the metabolic activities of anaerobic microorganisms and microbial metabolites (the content of EPS and SMP), and sludge particle size increases, resulting in serious membrane fouling [62]. The change of TMP is an important indicator that directly reflects the membrane fouling of an AnMBR, and the influence of HRT on TMP is shown in Figure 5.

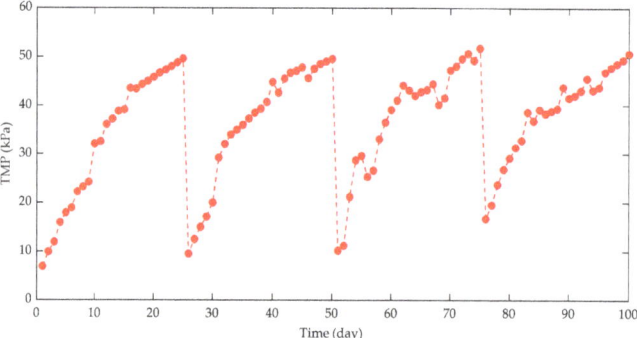

Figure 5. TMP at different hydraulic retention times (HRT).

This indicates that shorter HRT will lead to rapid increase of TMP and aggravate membrane fouling, which is not conducive to long-term stable operation of AnMBR.

Sludge properties have an important effect on membrane fouling [63]. The rate of membrane flux decline is positively correlated with the ratio of protein to polysaccharide, sludge settling performance and relative hydrophobicity, but negatively correlated with EPS [64]. With the extension of SRT, the total amount of EPS in the mixture showed a decreasing trend. The ratio of protein to polysaccharide in tightly bound extracellular polymeric substances increased, the sedimentation performance of sludge became worse, the relative hydrophobicity increased, the rate of membrane flux decreased, and membrane fouling aggravated [65]. At the same time, a large number of studies have shown that with the appropriate extension of SRT, MLSS increases, SMP concentration decreases, and membrane fouling is somewhat relieved. Therefore, the MBR system has an optimal SRT value.

3.2.4. Effect of Temperature on Membrane Fouling

Temperature changes affect the enzyme activity, mass transfer rate and microbial activity of anaerobic microorganisms. Changing the viscosity of the liquid can affect the treatment efficiency and stability of the reactor [66]. A study on the effect of temperature on the treatment effect and membrane fouling of the anaerobic membrane bioreactor [67] found that the microorganisms in the anaerobic membrane bioreactor can maintain high activity and COD removal rate at higher temperatures. In addition, the temperature will significantly affect the metabolism of microorganisms, resulting in different amounts of EPS secretion. At low temperatures, microorganisms secrete more polysaccharides and proteins for self-protection, resulting in higher EPS [68]. This information can help to optimize the anaerobic membrane bioreactor. The operating temperature should be found such that the system EPS concentration is at a lower level and therefore slowing down the fouling of the membrane [69].

Generally, the increase in temperature will decrease the viscosity of the mixed liquor in the MBR, (i) increase the solubility of suspended particles, (ii) increase the mass transfer diffusion coefficient [70], (iii) promote the movement of solute on the membrane surface to the bulk solution, (iv) reduce the thickness of concentration polarization layer, so as to improve the cross-flow velocity, and (v) increase the flux of the membrane [71].

When studying a forward osmosis membrane bioreactor, temperature has a significant effect on the forward osmosis process. The water flux increases with increasing temperature. The influence of the temperature of the draw solution is significantly higher than that of the feed solution, and only the temperature of the draw solution is increased, which can obtain higher water flux and lower membrane fouling under the premise of significantly reducing the heat consumption of the system, which is an efficient operation mode [72].

3.2.5. Effect of the Mode of Operation on Membrane Fouling

Among the many factors affecting membrane flux, the mode of operation is the key to the successful application of membrane technology in wastewater treatment. The cross-flow velocity can control the size of the permeate flux and to some extent contain membrane fouling [73]. The dynamic fouling layer formed by the pollutants on the surface of the membrane is the main reason for the decrease in membrane flux and the cross-flow filtration can effectively reduce the thickness of the dynamic fouling layer [74]. In cross-flow filtration, the addition of a gas such as nitrogen at the inlet end of the membrane module can increase the degree of turbulence in the pores and thereby enhance the shear-carrying effect on the fouling layer on the membrane surface, which will increase the flux. Increasing the cross-flow velocity helps to increase the permeate flux. However, as the cross-flow velocity increases, the increase in permeate flux decreases. Therefore, there is an optimal cross-flow velocity under a certain operating pressure, and the fouling is intensified above the optimal cross-flow velocity, resulting in flux attenuation [75].

For pressure-driven membrane filtration processes, operating pressure is the most direct factor. There is a critical operating pressure during operation. Exceeding the critical pressure causes the membrane to foul extremely seriously [76]. At higher operating pressures, the fine particles in the wastewater collect toward the inner surface of the membrane under high pressure, and the velocity is faster than the speed of the particles leaving the membrane surface, and particles form a contaminated layer on the membrane surface, causing rapid decay of flux. In addition, the increase in air volume is not proportional to the increase in flux [77]. As the amount of compressed air increases, the permeate flux does not increase and decrease. Therefore, it is also important to select the proper amount of air. On the one hand, it can reduce the fouling of the membrane surface. Increasing the permeate flux, on the other hand, does not cause unnecessary waste of energy consumption [78].

In addition, the gas/water two-phase flow can increase the degree of turbulence in the pores, thereby enhancing the shear-carrying effect on the fouling layer on the membrane surface, and effectively suppressing membrane fouling, thereby increasing the permeate flux [79]. Intermittent pumping is also an effective measure to delay the development of membrane fouling [14]. When the system is off, the TMP becomes zero and the back diffusion rate of foulants on the membrane surface is accelerated; the removal of back-diffused foulants near the membrane surface is also enhanced, and the membrane fouling is alleviated. However, the stoppage time cannot be too long. After stopping time reaches a certain level, the removal effect of membrane fluid scouring on membrane fouling will be greatly weakened, and the excessive pumping time will greatly reduce the system's water production, so it should determine the optimal pumping time ratio based on actual conditions [80].

3.3. Effect of Character of Activated Sludge Mixture on Membrane Fouling

The activated sludge characteristics affecting membrane fouling include sludge components, MLSS, sludge viscosity, environmental conditions (pH, DO), EPS, SMP, inorganic matter and microbial communities present in the activated sludge [81]. Excessive growth of activated sludge, especially overgrowth of filamentous bacteria, can lead to serious membrane fouling, which makes the frequency of membrane chemical cleaning higher and increases operating costs.

3.3.1. Effect of Activated Sludge Components on Membrane Fouling

The sludge mixture consists of three components, namely suspended solids, colloids and dissolved matter. Membrane fouling is the result of a combination of all the above three components [82]. The sludge mixture of an anaerobic membrane bioreactor has small particles and the resistance formed by the suspended solids of the sludge mixture accounts for 70% of the total resistance, the resistance due to the colloidal substance accounts for 22%, and the resistance of the dissolved substance accounts for 8% [83]. For an anaerobic membrane bioreactor, it is advisable to adopt membrane fouling control measures such as increasing the air used for mixing to increase the flushing strength of the membrane surface and adding coagulants to increase the suspended solids from the viewpoint of reducing the suspended solids resistance in the sludge mixture [84]. Membrane fouling control measures could also optimize the size of particles in the activated sludge. The relationship between sludge components and membrane fouling under different operating conditions was measured as shown in Figure 6.

Some studies [85] summarized the contribution of different sludge components to membrane fouling in MBR operations. Obviously, there are major differences between the research results, which can be attributed mainly to the following aspects: matrix conditions, membrane filtration performance, hydraulic conditions, SRT, biological state and component separation methods. The sum of the three kinds of filtration resistance calculated after the component separation of the mixture is usually greater than the direct filtration resistance of the mixture, which is the result of the fact that direct filtration of the mixture is conducive to the formation of dynamic layer and the development of reducing membrane fouling.

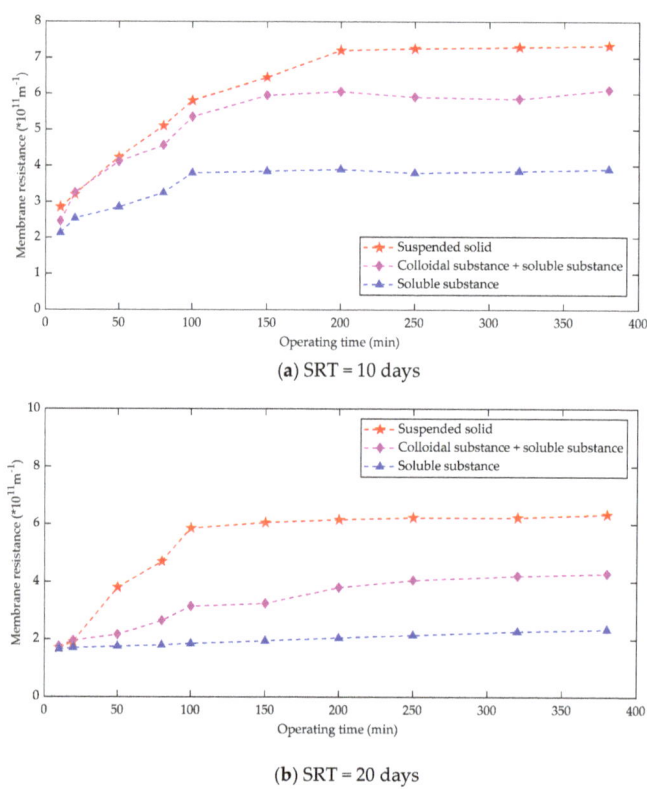

(a) SRT = 10 days

(b) SRT = 20 days

Figure 6. *Cont.*

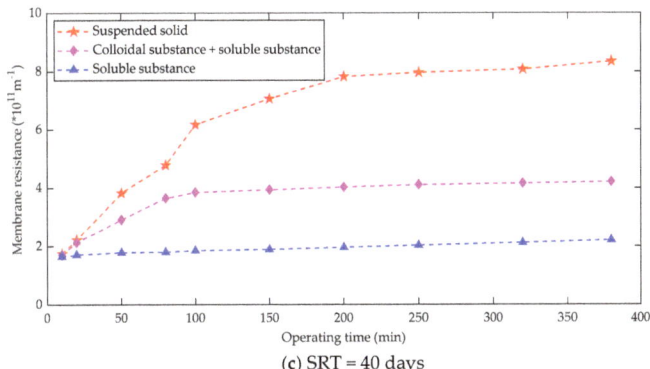

(c) SRT = 40 days

Figure 6. Relation between sludge composition and membrane fouling. (**a**) SRT = 10 days; (**b**) SRT = 20 days; (**c**) SRT = 40 days.

In the membrane fouling process of an integrated membrane bioreactor, a mud cake layer will be formed on the surface of the membrane, the resistance will increase, and the membrane flux will decrease. The formed mud-cake-layer resistance is related to the particle size. The smaller the particle size, the smaller the porosity of the mud cake layer, and the greater the resistance; the larger the particle size, the larger the porosity of the mud cake layer and the smaller the resistance of the mud cake layer [86]. When wastewater is treated anaerobically, microorganisms rely mainly on secreted extracellular enzymes to decompose macromolecular proteins, fats and polysaccharides into small molecules, and then degrade small molecules to produce CH_4 and CO_2. The sludge formed is relatively loose, and the size of suspended particles in the mixed liquor is small [66].

3.3.2. Effect of MLSS on Membrane Fouling

In the application of a membrane bioreactor, MLSS is an important process parameter which directly affects the performance of the membrane [46]. On the one hand, the high MLSS concentrations can reduce the sludge loading rate, improve the treatment efficiency and increase the viscosity of the mixed liquor; on the other hand, it can result in the increase in membrane filtration resistance. At the same time, the increase of sludge mass concentration will cause hypoxia or anaerobic phenomenon at the bottom of the reactor. In addition, the sludge concentration will increase and the aeration amount will remain unchanged, resulting in an anoxic state inside the activated sludge and short-range nitrification and denitrification in the reactor, so that the total nitrogen has a good removal rate [87].

The sludge mass concentration gradually increases with the increase of sludge age. At the beginning of the operation of an MBR, the sludge mass concentration increases greatly, mainly because of the strong microbial metabolism and high sludge load during this period. After that, the sludge growth is slow but the sludge mass concentration increases rapidly until the middle stages of operation. In the later stage, the sludge mass concentration is limited by organic pollutants, and after the sludge mass concentration reaches a certain level, it no longer grows and gradually reaches a stable state [88]. When the sludge mass concentration is stable, the COD of the supernatant and effluent will fluctuate with the COD in the raw wastewater, but the total removal rate of all the solutions' COD and removal rate of the supernatant COD will no longer show significant fluctuations. It shows that the system is in stable state and has good removal effect on organic pollutants [89].

Under the condition that the flow rate of the water is kept constant, as the mass concentration of the sludge increases, the viscosity of the mixed liquid also increases, causing serious membrane blockage, resulting in a decrease in membrane porosity, thereby increasing filtration resistance and increasing TMP [90]. Studies have shown that the higher the sludge mass concentration, the greater the membrane filtration resistance. Although higher MLSS can increase the volumetric load of the membrane bioreactor, the increase of membrane resistance will increase energy consumption and therefore the operating costs, and affect effluent quality. Thus, the MLSS of the membrane bioreactor should not be too high, and should be considered in terms of treatment efficiency and processing capacity [33].

3.3.3. Effect of Sludge Viscosity on Membrane Fouling

Viscosity is essentially the ability of a molecule or solid particle in a liquid to resist external stress or shear forces. The greater the viscosity of a solution, the greater its ability to withstand external stresses or shear forces [49]. The mixed liquor contains a large amount of viscous substances such as EPS, which makes it easy for the sludge flocs to adhere to the surface of the membrane, thereby accelerating membrane fouling, and reducing the gas-liquid flow rate generated by the aeration as well as forming a shearing effect on the membrane surface. This slows down the erosion of the contaminants on the membrane surface and worsens the operation of the membrane bioreactor [91].

Excessive viscosity of the sludge mixture increases the likelihood of sludge adhering to the membrane surface, thereby accelerating membrane fouling. In addition, Hu [92] showed that the sludge with high viscosity is not easy to clean after being adsorbed to the surface of the membrane, resulting in poor recovery of membrane flux.

3.3.4. Effect of EPS and SMP on Membrane Fouling

The effects of EPS and SMP on membrane fouling have received increasing attention in recent years. Figure 7 is a representation of the relationship between EPS, blend EPS (BEPS), SMP, active units (bacterial micelles and biofilm) and ECMs.

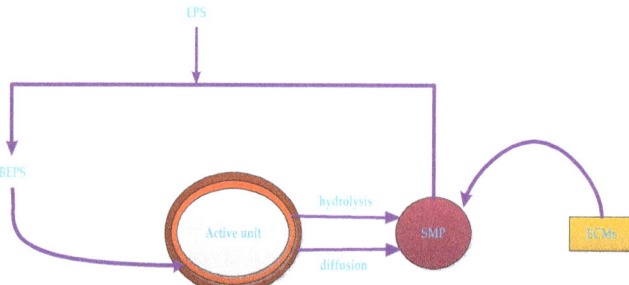

Figure 7. The relationship between EPS (extracellular polymers), BEPS (fixed EPS), SMP (soluble microbial product) active units and ECMs.

EPS includes insoluble organic matter secreted by cells, which is shed from the cell surface or caused by cell death. The main components are protein (EPSp) and carbohydrate (EPSc) [93]. The EPS in the sludge mixture and sludge floc is usually extracted by heating, organic solvent extraction and ion exchange. The EPS is also dissolved in the mixed liquor. It is a soluble microbial metabolite termed as soluble microbial product (SMP). EPS and SMP levels are typically characterized by CODcr, TOC, UV254 or directly by protein, polysaccharide and humic acid content [94].

EPS is an important factor affecting sludge settling performance and membrane fouling rate in MBR. Excessive EPS will further deteriorate the mutual flocculation effect between microorganisms and weaken microbial flocs [95]. The difference in EPS concentration of different activated sludge reflects the difference in the filtration performance of the corresponding sludge. The capillary sunction time (CST) value of the normal sludge is significantly smaller than that of the sludge bulking, which means that the filtration performance is good when the EPS concentration is low [96]. On the other hand, due to the high P/C ratio and hydrophobicity, the expanded sludge containing EPS is easily adhered to the surface of the membrane, thereby causing membrane resistance. At the same time, as the adsorption time prolonged, the membrane flux decreased significantly, indicating that the adsorption of EPS on the membrane surface will lead to irreversible fouling of the membrane [97]. Furthermore, EPS is a key factor affecting sludge agglomeration. The decrease of EPS causes the aggregation of sludge flocs to decrease, so that the size of sludge flocs becomes smaller, and smaller size flocs are easily deposited on the membrane surface. This will cause the fouling of membrane. Therefore, there should be an optimized EPS concentration in the membrane bioreactor, at which the floc structure of the sludge can be maintained, and the membrane fouling potential of the sludge flocs is minimized [98].

SMP is a large class of soluble organic matter produced by microbial metabolism, including polysaccharides, proteins, humic acids and nucleic acids. The composition is extremely complex and poor in biodegradability [99]. Due to the entrapment of the membrane, the adsorbed SMP accumulates on the surface of the membrane to cause concentration polarization, thereby causing membrane fouling. The formation of sedimentary layers is a major factor in membrane fouling [100]. During the formation of the sedimentary layer, SMP continuously fills the gap of the microbial flocs, making the deposited layer more dense, resulting in a decrease in the porosity of the deposited layer, resulting in a decrease in permeability and an increase in the specific resistance of the deposited layer. The higher the concentration of SMP in the sludge mixture, the denser the structure of the sediment layer and the smaller the void fraction. As the sedimentary layer continues to develop, the membrane flux decreases, which increases membrane fouling [101].

3.3.5. The Effect of Microorganisms on Membrane Fouling

The biological phase in the MBR will change the sludge morphology, particle size distribution (PSD), EPS, viscosity and other parameters affecting membrane fouling. Compared with the traditional activated sludge method, the evolution of microbial populations in the MBR system is characterized by few species, which are vast and with obvious dominant populations. With the extension of the running time, the dominant populations in the MBR system consist of swimming ciliates, worms and bell worms, beetles, and red spotted worms. Membrane fouling can be predicted by dominant populations that indicate the state of the sludge [102].

The difference in biophase has a large effect on membrane filtration resistance. When the biophase is changed from a lower protozoan to a higher protozoa, it is reflected as a decrease in the starting point of the filtration resistance and slowing point in the growth rate of the microbial community. Wang et al. [103] showed that the micro-animals and activated sludge in MBR are a dynamic process of interaction. Meng et al. [104] found that filamentous bacteria play an extremely important role in membrane fouling during the operation of MBR. Excessive or too few filamentous bacteria can cause serious membrane fouling. The sludge flocs lacking filamentous bacteria are relatively fine, and it is easy to cause serious membrane pore blockage, and the activated sludge containing excessive filamentous bacteria will form a thick and firm filter cake layer, which increases the filtration resistance [105].

4. Membrane Fouling Control

4.1. Modification of Membrane Material Body

4.1.1. Physical Blending

The physical blending is to physically mix the membrane material matrix with the modifying additive in a certain ratio, and the modified additive does not react with the bulk of the membrane material [106]. Physical blending modifications can balance the advantages and characteristics of the bulk membrane material and additives and have the advantages of both obtaining better cast film materials [107]. At present, hydrophilic materials blended with PVDF membrane materials can be roughly classified into two types. One is a hydrophilic polymer material, and the other is a small molecule inorganic particle. However, physical blending of membrane materials also has difficulties such as poor compatibility [108].

The advantage of blending modified hydrophilic polymer with PVDF polymer is that after adding relevant hydrophilic polymer to the film material, it can compensate for various performance defects of the film with raw material alone, and also can give the membrane material itself the new superiority [109]. In the polymer blend membrane, the compatibility between the polymers has a direct influence on the formation and structure during the phase separation process of membrane synthesis [110].

At present, in the improvement of the performance of the separation membrane, the polymers reported for blending with PVDF including polyvinylpyrrolidone (PVP), chloromethylated polysulfone (CMPS), polymethyl methacrylate (PMMA), poly vinyl acetate (PVAc), PEG, polyvinyl alcohol (PVA), sulfonated polystyrene (SPS), nylon 6, sulfonated polysulfone, polyacrylonitrile (PAN), polysulfone, and sulfonated polyaryl ether sulfone (SPES-C) [111].

Compared with hydrophilic polymer blending, blending small-molecular inorganic particles with PVDF to improve the hydrophilicity of the membrane is a rapid development in the recent years [112]. Commonly used inorganic particles are Al_2O_3, SiO_2, TiO_2, etc. The modified membrane prepared by using the blend casting solution perfectly combines the properties of the PVDF membrane with the hydrophilicity and heat resistance of the inorganic material to form a novel organic–inorganic composite membrane [113].

4.1.2. Chemical Copolymerization

The chemical copolymerization modification can increase the hydrophilicity of the organic polymer film. This is carried out by adding a modified monomer to the original cast film material of the organic polymer film to cause a complicated copolymerization reaction and to form a new copolymer as a cast film material. Common methods for modifying membrane materials include copolymerization, block, chain extension, grafting, and so on. Sun et al. [114] induced the copolymerization of sulfonamide amphiphilic groups with acrylonitrile materials and modified the propylene amine groups to inhibit the protein adsorption properties and modified the polypropylene-sulfonamide copolymer membrane materials with excellent properties. The PVDF was modified with 10% NaOH to improve the hydrophilicity of the ultrafiltration membrane, and the hydrophilicity of the composite membrane obtained by grafting polyoxyethylene methacrylate (POEM) with 5 wt % to PVDF was also greatly improved [115].

Selina et al. [116] studied the factors that affect the performance and treatment efficiency of direct membrane filtration, and pointed out that membrane fouling is the main challenge of direct membrane filtration. Direct membrane filtration has been used as a promising technology for wastewater recovery and resource recovery in various laboratories and pilot scale studies, which is attributed to the advantages of direct membrane filtration process [117]. For example: (i) Direct membrane filtration processes have a relatively simple system configuration, requiring less capital cost and footprint. (ii) It has been well documented that direct membrane filtration of wastewater treatment could produce superior permeate quality that meets water discharge or reuse standards and effective concentration of nutrients for further recovery [118]. (iii) In some direct membrane filtration processes, water reclamation and resource recovery from wastewater can be simultaneously achieved, showing great potential for saving energy consumption, improving carbon neutrality, and minimizing footprint compared to conventional wastewater treatment processes [119].

4.2. Hydrophilic Modification of the Surface of Membrane Material

4.2.1. Surface Coating

The surface coating modification method refers to coating a hydrophilic substance on the surface of the hydrophobic film to be modified, thereby improving the anti-fouling performance of the film. A comb polymer was coated on a PSF ultrafiltration membrane and the modified membrane was characterized by infrared spectroscopy and X-ray photoelectron spectroscopy (XPS). The presence of the coating layer was confirmed by the above characterizations [120]. Furthermore, it was found through experiments that the flux recovery rate of the modified membrane was significantly improved. In addition, when hydrophilic materials were coated on the surface of PVDF ultrafiltration membrane, it was found that the flux recovery rate of the modified membrane was more than 90% on the basis of improving the anti-fouling performance. Although the modification method is relatively simple in operation, the modified molecules coated on the film are easily detached from the surface of the film, and therefore long-term stable modification was unable to be achieved. Zhao et al. [121] first immersed the plasma-pretreated polyvinylidene fluoride (PVDF) membrane material in a 0.4% solution of TMC-hexane to form a modified PVDF-TMC membrane, and then immersed the modified membrane in sequence. The self-assembled coating modified layer was formed on the surface of the film with 0.5 wt % of SiO_2-NH_2 nanoparticle suspension and 0.1 wt % of SA solution. The experimental results showed that this method can effectively improve the anti-fouling performance [122].

4.2.2. Membrane Modification by Low Temperature Plasma Surface Treatment

The plasma modification methods such as only plasma modify treatment (OPMT), plasma polymerization graft coating treatment (PPG-CT) and plasma trigger radical graft polymerization (PTRGP) on the surface of polymer materials can modify the membrane, but the modification mechanisms are slightly different. PTRGP method is the grafting of monomer on the surface of the membrane [123]. The copolymerization is mainly characterized by the fact that the monomer can penetrate into the pores of the membrane to carry out the grafting reaction, but the level of grafting is small. PPG-CT is mainly composed of the polymerization of monomer radicals, and its function is to provide high grafting rate. The morphology of the membrane surface changes greatly. The presence of polymer is obvious from the scanning electron micrograph of the membrane surface. OPMT is mainly caused by plasma ion incidence, plasma etching and plasma surface crosslinking. At high temperatures, the loss of membrane quality was more serious [124].

The low-temperature plasma treatment modification technology uses a gaseous substance such as an atom, a molecule or an ion in a state in which the positive and negative charges are in a plasma state to attack the polymer, and induces a chemical reaction such as hydrogen elimination on the surface of the membrane to introduce a large number of polar groups such as -OH, -COOH, -SO_3H, -CO, and -NH_2 [125]. This causes grafting of other monomeric substances on the surface of the membrane material to change its properties. The low-temperature plasma treatment modification technology can be realized in a medium to low temperature environment without changing the excellent characteristics of the original membrane. However, the instruments and equipment required for plasma treatment are generally costly and the operating conditions are harsh, thus limiting its use [126].

4.2.3. Surface Grafting

Surface grafting refers to the modification of the thin layer on the surface of the membrane without changing the bulk properties of the membrane material so that a relatively stable chemical bond is formed between the membrane surface and the grafted polymer chain [127]. The effect of modification is more durable. Zhan et al. [128] prepared a CMPSF microporous membrane by phase inversion method, and introduced a large amount of primary amine groups on the surface of the membrane by chemical modification, thus constructing a surface initiation system -NH_2/$S_2O_8^{2-}$. The monomeric DMAEMA was successfully graft polymerized on the surface of the polysulfone microfiltration membrane to form a porous graft membrane PSF-g-PDMAEMA. Such a membrane can remove CrO_4^{2-} ions in the water effectively by selective adsorption.

In recent years, research on membrane fouling control methods related to irradiation grafting has attracted more and more attention [129–133]. The reaction mechanism of UV irradiation grafting belongs to free radical reaction, which has low polymerization temperature and mild polymerization conditions. Some researchers have used ultraviolet radiation grafting to introduce acrylic acid into the polypropylene film, which greatly improved the hydrophilicity of the polypropylene film [129]. In the same way, the thio betaine methyl acrylate monomer was grafted onto the polypropylene microporous membrane, and the contact angle of the membrane was significantly reduced, and therefore the flux increased significantly [130]. In addition, the continuous hydrolysis of ultraviolet radiation was used to graft the pentaerythritol monoester to the surface of the polypropylene hollow fiber membrane, and the hydrophilicity and anti-fouling of the membrane were improved significantly [131]. It is also possible to graft hydroxyethyl methacrylate onto the surface of the polypropylene microporous membrane by ultraviolet light irradiation by using benzophenone and ferric chloride as co-initiators, so that the contact angle of the modified membrane is reduced [132].

Jin et al. [133] used polypropylene hollow fiber membrane as the matrix and sodium styrene sulfonate as the hydrophilic monomer to prepare a modified hollow fiber membrane with a different grafting rate under ultraviolet irradiation and infrared spectrum. The electron microscopy test proved that the modification was successful. When the graft ratio was 13.3%, the water contact angle of the film was 46°, which was significantly lower than that of the original film (64°) [134]. Up to a certain value, as the grafting rate was increased, the water flux increased. However, as the grafting rate increased further, the water flux decreased. When the grafting ratio was 9.0%, the water flux reached a maximum of 102 L/(m^2·h), an increase of 29 L/(m^2·h) compared to the original film. The hydrophilicity of the polypropylene hollow fiber membrane modified by ultraviolet irradiation improved significantly.

4.3. Optimization of Membrane Modules

Factors that should be considered in the optimization of the membrane module are the shape of the membrane module, the placement of the module, hydraulic conditions, the diameter and the length as well as the tightness of the hollow fiber filaments [135]. As can be seen from Figure 8, the wall shear stress inside the membrane component decreased with the increase in the length of the filament.

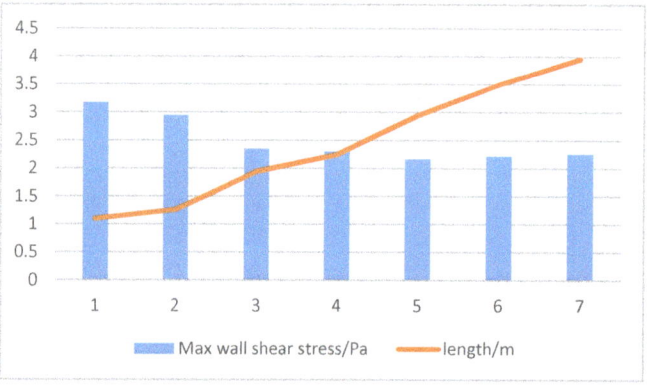

Figure 8. Effect of membrane module length on wall shear stress.

Xiong et al. [136] used the Euler model and the porous medium model to calculate the fluid flow in the membrane module with different structures. The calculation shows that reducing the diameter of the aeration hole and increasing the number of aeration holes can promote the uniform distribution of gas–liquid two-phase flow field and liquid phase velocity field, as well as wall shear stress and turbulent viscosity enhancement [137]. Increasing the height of the membrane module is beneficial to increase the membrane area of a single membrane module while making full use of the gas for scrubbing during aeration. The liquid two-phase flow performs high-efficiency air scrubbing on the wall surface of the membrane. Xu [138] designed a novel spiral membrane module with a certain rotation angle by bionics principle, thereby improving the sensitivity of the membrane module to water, gas and other fluid disturbances, resulting in vibration of the membrane and increasing the elastic collision between the bubble and the membrane surface [139]. The concentration polarization phenomenon in the falling film separation effectively controls the membrane fouling, improves the membrane separation efficiency and flux, and reduces the energy consumption [140]. A Box–Behnken method can be used to optimize hydraulics of the membrane module. Multi-parameters such as inlet diameter, inlet length, membrane shell height, inlet/outlet end tube length, billet structure diameter, and inlet and outlet tilt angle can be used as variables in the experimental design. The Box–Behnken method is the synthesis of statistical design experiment technology. It uses the design of the Box–Behnken experiments and obtains data through the experiments to find the proper multiple quadratic equation which can fit the functional relationship between the factors and the effect value well. The optimal process parameters can be determined through the analysis of the regression analysis. The Box–Behnken method solves the multi-variable problem as a statistical method [141]. A membrane module design with optimal response variables can obtained by a series of experiments with varying membrane configurations. The particle residence time distribution and hydrodynamic characteristics of the liquid–solid two-phase flow in the three-dimensional model can be simulated by coupling the calculation between the Reynolds stress RSM turbulence model and the discrete phase model (DPM) based on the Euler–Lagrange algorithm [142]. The simulation results show that the velocity distribution of the shell surface of the cyclone-enhanced membrane module is more uniform and the shear stress of the membrane surface is high; the turbulent dissipation rate and vorticity distribution are different from those of the traditional membrane module [143]. The experimental results confirmed that the optimized membrane module has the characteristics of high yield of flux, low pressure drop and low membrane fouling rate [144]. To overcome the shortcomings of flat sheet membranes, a new type of folding membrane module has been designed. The vertical inclination and membrane spacing of the folding membrane module were optimized by constant pressure membrane filtration experiments, which greatly increased the maximum steady state membrane flux and the rate of transmembrane pressure rise. The reduction of the rate of transmembrane pressure rise is normal, and the removal rate of chemical oxygen demand (COD) and NH_3-N is also improved [145]. Viet et al. found that, compared with the traditional membrane bioreactor, the osmotic membrane bioreactor (OMBR) has a broader application prospect in reducing membrane fouling and improving effluent quality [146]. Blandin et al. thought that the OMBR process is expected to consume less energy than MBR process, but further research is needed to confirm this [147].

4.4. Changing the Properties of the Feed Water

There are some factors affecting membrane fouling that exist in the feed liquid. In general, the direct method of controlling the characteristics of the feed liquid is to add a flocculant or adsorbent to the feed liquid. Compared with the monomer salt, the polymeric salt can provide more positive charge and electrically neutralize the suspended particles in the liquid, which can improve the removal rate of the suspended particles and increase the diameter of the particles [148]. The indirect method is to control the reaction by changing the sedimentation and flocculation performance of the feed liquid by adjusting the operating conditions such as HRT, SRT and reaction temperature.

Adding a commonly used adsorbent such as powdered activated carbon (PAC) to MBR can effectively control the development of membrane fouling, slow the rate of increase of TMP, and prolong the membrane operating cycle [149]. PAC has high adsorption capacity and can absorb dissolved organic matter, EPS, microparticles, etc. in the mixture, and can also be embedded in the sludge as a skeleton to form more solid sludge particles, which are not easily damaged by shear force and therefore will not release pollutants back into the bulk liquid. However, as the system operates, the PAC will gradually saturate, requiring regular replacement; therefore an optimum value for the dosage of PAC exists. Zeolite can also be an effective adsorbent. Zeolite can adsorb a part of sludge particles and reduce the resistance to membrane filtration [150]. However, zeolite also has an optimum dosage. Excessive zeolite will be adsorbed onto the surface of the membrane and increase the filtration resistance. The tiny particles brought along by the zeolite itself will increase the degree of membrane clogging and increase membrane fouling.

4.5. Control of Operating Conditions

The main operating conditions of MBR are membrane flux, operating pressure, aeration, pumping time and cleaning cycle. One measure commonly used in MBR is to control the membrane flux below membrane critical flux or sustainable flux operation. Zhang et al. [151] studied the short-term and long-term actual operation of the small-scale immersed MBR (SMBR) under constant flux and constant pressure modes of operation and found the constant pressure operation in the subcritical region and the constant pressure below the economic operating pressure TMP in the subcritical region was conducive to the long-term stable operation of the MBR. This is beneficial to the long-term stable operation of the MBR [152]. Membrane flux and membrane resistance have a great relationship with the shear force of membrane caused by the gas flow. In a certain range, the membrane flux will increase with the increase of aeration. When the aeration amount reaches a threshold, the flux will remain the same, and with further increase in the aeration flow rate, the membrane flux will decrease. This is because an excessive amount of aeration will break up the already flocculated suspended particles, making the particle size of the suspended particles smaller and more likely to block the pores of the membrane. Intermittent suction is also an effective measure to control membrane fouling. Zhang et al. [153] considered that the cleaning cycle and use time of the membrane had little effect on membrane fouling. In addition, ultrasonic irradiation can play a role in slowing down membrane fouling. Studies have shown that ultrasound has the effect of mitigating membrane fouling and can extend the membrane cleaning cycle [154]. The addition of ozone to the feed liquid can also effectively reduce membrane fouling. Wu et al. [155] showed that the optimum dosage (O_3/SS) is 0.25 mg/g per day, which can reduce the EPS in the supernatant and enhance the suspended solids. In general, the control measures for membrane fouling are based on the factors that control membrane fouling. The common methods and principles of controlling membrane fouling are shown in Table 3. It is worth noting that the effect of a single control method is not ideal. In the actual application process, various methods should be combined according to the specific conditions to achieve the desired effect [156].

4.6. Cleaning of Membrane Fouling

Membrane cleaning can effectively remove and control membrane fouling, reduce TMP, and restore membrane flux [71]. The cleaning methods of contaminants in the MBR process can be divided into physical cleaning, chemical cleaning, and electric cleaning. In the actual operation process, it is difficult to achieve the best results with a single cleaning method. Generally, a combination of several methods will be effective in cleaning the membrane [157].

4.6.1. Physical Cleaning

Physical cleaning mainly removes reversible contaminants in the membrane surface or membrane pores and the methods mainly include aeration, backwashing (air or filtrate), ultrasonication, sponge scrubbing and water washing. Physical cleaning allows the MBR to operate at a relatively constant flux without causing secondary contamination, but requires frequent cleaning at increased operating costs [47]. Aeration is the most commonly used membrane cleaning method in aerobic SMBR. It uses the cross-flow caused by the ascending airflow to reduce the deposition of particles on the membrane surface and flushes the membrane surface pollutants to reduce membrane fouling. In the operation of the hollow fiber membrane process, the role of aeration is to provide the oxygen demand for degrading organic matter, supply the oxygen demand for the growth and metabolism of the activated sludge itself, and remove contaminated deposits on the membrane surface [25]. A large amount of aeration produces a severe turbulent state on the membrane surface, and a strong shear force can carry away the filter cake layer deposited on the membrane surface. Intermittent operation and aeration combined can enhance the dispersion of pollutants attached onto the membrane surface and effectively retard membrane fouling. The suction is stopped after the membrane fouling occurs, and the continuous aeration of the membrane cake layer from the membrane surface can restore the flat sheet membrane flux [158]. However, there is also an optimal threshold for aeration.

Backwashing can remove most of the reversible pollutants and improve the membrane filtration performance. The key parameters affecting the backwashing effect are the backwash intensity, frequency, time and frequency-time ratio. Jiang et al. [159] found that low frequency, long time backwashing (600 s filtration/45 s backwashing) is more effective than high frequency, short time backwashing (200 s filtration/15 s backwashing). Fan et al. [160] obtained experimental methods and theoretical derivation for determining the optimal backwashing cycle of MBR, which can be used for automatic control of MBR backwashing. In the actual operation of MBR, cleaning by aeration and backwash are usually used in combination, which can achieve better results than a single method of cleaning.

Ultrasonic online cleaning methods can effectively control membrane fouling [13]. Ultrasonic waves have very special properties. The "cavitation" caused by the instantaneous release of ultrasonic energy concentrates on the solid–liquid interface, thus exerting a strong impact on the point of action and its surroundings and on the gel layer attached to the surface of the membrane. The precipitate produces a strong peeling effect [161]. However, excessive ultrasonic strength and time of action will break up the sludge flocs, affecting sludge activity and damaging the membrane module [99]. Therefore, the selection of appropriate ultrasonic intensity and time of action is essential for effective control of membrane fouling.

Table 3. Comparison of membrane fouling control methods.

Control Methods	Controlling Factors	Expected Results	Precautions
Modification of membrane material	Improve membrane surface hydrophilicity	Reduce the adsorption of impurities on the membrane surface and membrane pores	The membrane material should be modified according to treatment objectives
Optimization of membrane components	Improve membrane surface water conditions	Improve the effect of membrane surface gas flow flushing and decontamination	High mechanical properties for membrane materials
Aeration, ultrasound	Remove membrane deposits and improve liquid properties	Gas–liquid flow flushes out membrane deposits to increase activated sludge activity	Excessive aeration or microwave vibration will break up the sludge flocs and increase the fouling of the membrane
Add flocculant or adsorbent (PAC), ozone	Improve liquid properties	Improve sludge settling and reduce EPS and SMP in feed liquid	Inorganic flocculants change the pH of the feed, the adsorbent itself may also become a contaminant, and ozone inhibits microbial activity
Intermittent suction	Improve film surface detachment properties	Conducive to the membrane surface gas flow flushing with pollutants	Too long stoppage will affect the amount of water produced, too short to achieve the desired results

4.6.2. Chemical Cleaning

Chemical cleaning is required when physical cleaning does not meet membrane fouling requirements. Commonly used chemical agents include alkali cleaning agents, acid cleaning agents, oxidizing cleaning agents, and surfactants (ethylenediaminetetraacetic acid EDTA, ammonium hydrogen fluoride, etc.) [162]. Alkali cleaning agents can effectively remove organic matter and biological foulants [163]. The process is as follows: inject water into the cleaning water tank, heat it with steam, start the cleaning pump, slowly add the cleaning agent, mix to make the cleaning agent completely dissolved, first clean the first section, then clean the second section for dynamic circulation for 40 min, and then soak for 50 min to clean alternately. When the pH is reduced by 0.5, add NaOH to control the pH value at 10–11. When the pH value is no longer reduced, carry out water washing. When the pH value of the effluent reaches 6–7, the water washing is finished [164]. An acid cleaning agent can effectively remove mineral and inorganic fouling [165]. During acid cleaning, water is injected into the cleaning water tank, heated by steam, the cleaning pump is started, hydrochloric acid is added slowly, the pH value is controlled at 2–3, and the cleaning is carried out in sections. The first section is cleaned, and then the second section is cleaned. The dynamic circulation is 40 min, and then the immersion is 40 min. In this way, the cleaning is carried out alternately. When the pH value is no longer increased, the water is washed [166]. When the pH value of the water reaches 6–7, the water washing is finished. Hydrochloric acid can remove hydrophobic organics better, while sodium hydroxide can remove more organic pollutants. The combination of the two can effectively remove the pollutants on the membrane surface, but the removal effect on the pollutants inside the membrane pore is poor [167]. Oxidizing cleaning agents can increase the hydrophilicity of organic polymer contaminants and can effectively remove the adhesion in the pores of the membrane. A surfactant can improve the contact of the cleaning agent with the pollutants, improve the cleaning effect, and it can also destroy the bacterial cell wall and weaken the foulants caused by the biofilm. Chemical cleaning can be used for both on-line cleaning and off-line cleaning, which can greatly restore membrane flux, but the cleaning waste can sometimes cause secondary fouling [168]. The four major factors to consider in chemical cleaning are: concentration of the cleaning agent, cleaning temperature, contact time, and mechanical strength of the film [169].

4.6.3. Electric Cleaning

Electric cleaning achieves the effect of removing contaminants by applying an applied electric field on the film at a certain time interval to cause the contaminating particles to move away from the film surface in the direction of the electric field. However, this method requires that the film has a conductive function, or that the electrode can be mounted on the film surface, so that it is used less [170].

4.6.4. Ultrasonic Cleaning

Ultrasonic irradiation can clean the fouled membrane through the production of important physical phenomena including microjet, microstream and shock waves [171]. Indeed, the particles can be released from the fouled membrane by the aforementioned physical phenomena taking place in a heterogeneous liquid–solid interface. Furthermore, the active hydroxyl radicals generated in the presence of ultrasonic irradiation can attack the adsorbed foulants and degrade the molecules of foulants which consequently result in membrane fouling control [172]. However, the membrane can be damaged through chemical reactions between the generated hydroxyl radicals and the membrane [173]. Therefore, the operational conditions should be optimized in ultrasound-MBR hybrid systems. The ultrasonic cleaning can be performed either in situ (online) or ex situ (offline) for cleaning the membrane of MBRs. Moreover, pretreatment of the wastewater by ultrasonic irradiation or by hybrid ultrasound methods prior to MBRs can decrease the organic loading of the wastewater and subsequently postpone the fouling of the membrane. Moreover, the ultrasonic method can be combined with other cleaning methods, i.e., chemical cleaning and backwashing, to improve the cleaning efficiency [174].

5. Conclusions

MBR technology is a highly competitive technology and has been widely used in various fields of wastewater reclamation and wastewater recycling. However, membrane fouling is a hindrance to the widespread promotion of this technology. Therefore, research on the causes, mechanisms and control technologies of membrane fouling is of vital importance to this technology. Although some achievements have been made, there is room for more improvements. The author believes that future research should focus on the following aspects:

(1) Due to the heterogeneous structure and complexity of WOM or the dissimilar fouling behaviors between different organic surrogates, it is highly recommended that use should be made of real wastewater in combination with advanced DOM analysis such as FT-ICR-MS (Fourier Transform-Ion Cyclotron Resonance-Mass Spectrometry), SEC-OCD, and CLSM, as well as online monitoring methods

such as quartz crystal microbalance, or any visualization apparatus for exploring real-time organic membrane fouling formation. Real-time monitoring capabilities are also of great benefit for the optimization of the periodical cleaning of membranes during long-term FO processes. New data on the complex interactions between organic foulants and membrane materials can help the development of effective fouling control strategies for real wastewater. For instance, appropriate pretreatment can be developed/designed to effectively eliminate high molecular weight biopolymers (e.g., polysaccharides and proteins) to mitigate FO membrane fouling. Finally, considering the typical short-term formation of irreversible fouling in FO processes, the limited operational test period, and bench-scale nature of many previous studies, pilot-scale FO systems should be operated and studied with due provision for long-term monitoring if transition to successful full-scale wastewater FO processes is to be realized.

(2) An alternative solution for improvement of direct membrane filtration performance is to develop new membranes with increased anti-fouling properties. Several reported studies focused on developing the novel anti-fouling membranes and the lab-scale testing findings displayed their good performances in membrane fouling alleviation. In view of the absence of large-scale direct membrane filtration processes in the market, further research needs to be emphasized on (i) membrane fouling control technologies of direct membrane filtration, especially towards low energy consumption, less chemical usage, and easier operation and maintenance; (ii) development of novel membranes, especially having a mechanically robust nature with low-cost environmentally friendly materials and self-cleaning properties; (iii) comprehensive economic analysis, life cycle assessment, and carbon footprint analysis of different direct membrane filtration processes in order to identify the most suitable system configuration for further scale-up.

(3) The effect of the presence of ultrasound-active inorganic nanoparticles in the matrix of the membrane used in MBR systems can be investigated in future research. Ultrasound-active nanoparticles, i.e., ZnO, TiO_2, etc., can produce hydroxyl radicals in the presence of ultrasonic irradiation. Generated hydroxyl radicals can degrade the foulants adsorbed on the surface of the membrane or captured in pores of the membrane, which consequently results in controlling the membrane fouling. High energy consumption of the ultrasonic transducers limits the application of ultrasound-MBR systems on a full scale. Further research is needed on the hybrid methods with low energy consumption for improving the application of ultrasonic technology in full-scale MBR systems.

(4) Aeration optimization, such as intermittent or cyclic aeration, automatic aeration control based on DO- or nutrient removal feedback and mechanically-assisted aeration scouring, has attracted much attention to achieve efficient membrane fouling control with less energy consumption. The method of aeration could be further optimized according to the CFD modeling on the fluidization and the scouring behavior of the particles in MBRs. Moreover, the attachment tendency of biofilm colonizers on the medium and membranes should be assessed. Moreover, chemical cleaning efficiency is highly related to the interaction between chemicals and foulants. The chemical reagent has greater potential to decrease the aging of membranes and even lead to the inactivation of microorganisms in the bioreactors. These adverse effects caused by the chemical cleaning will be highlighted in the future.

Funding: This research was funded by the National Natural Science Foundation of China (No. 61563032), the Natural Science Foundation of Gansu Province (No. 1506RJZA104), the University Scientific Research Project of Gansu Province (No. 2015B-030), and the Excellent Young Teacher Project of Lanzhou University of Technology (No. Q201408).

Conflicts of Interest: The authors declare no conflict of interest.

Abbreviations

AnMBR	anaerobic membrane bioreactor
BEPS	blend extracellular polymer
BPC	biopolymer clusters
CFD	computational fluid dynamics
CFV	cross-flow velocity
CLSM	confocal laser scanning microscopy

CMPSF	chloromethylated polysulfone
COD	chemical oxygen demand
CST	capillary suction time
DO	dissolved oxygen
DMAEMA	dimethylaminoethyl methacrylate
DPM	discrete phase model
ECM	extracellular matrix
EDTA	ethylene diamine tetraacetic
EPS	extracellular polymers
F/M	food to microorganism ratio
FO	forward osmosis
HA	humicacid
HPI	hydrophilic
HRT	hydraulic retention time
HPO	hydrophobic
MBR	membrane bioreactor
MF	microfiltration
MLSS	mixed liquid suspended solids
OMBR	osmotic membrane bioreactor
OPMT	only plasma modify treatment;
OLR	organic loading rate
PAC	powdered activated carbon
PAN	polyacrylonitrile
pH	hydrogen ion concentration
PE	polyethylene
PES	polyethersulfone
POEM	polyoxyethylene methacrylate
PPG-CT	plasma polymerization graft coating treatment
PTRGP	plasma trigger radical graft polymerization
PVDF	polyvinylidene fluoride
PSD	particle size distribution
PVC	Polyvinyl chloride
PVP	Polyvinyl pyrrolidone
PS	polysulfone
PVAc	polyvinyl acetate
PVA	polyvinyl alcohol
SA	sodium alginate
SEC-OCD	size exclusion chromatography with organic carbon detector
SF-DMBR	self-forming dynamic membrane bioreactor
SMB	sponge-based moving bed
SMBR	small-scale immersed MBR
SMP	microbial metabolites products
SRT	sludge retention time
SG	suspended-growth
SPS	sulfonated polystyrene
SPES-C	sulfonated polyaryl ether sulfone
TMP	transmembrane pressure
TOC	total organic carbon
UV	under voltage
WOM	wastewater organic matter
XPS	X-ray photoelectron spectroscopy

Appendix A

Table A1 shows the review papers on membrane fouling in specific areas in recent years.

Table A1. Review papers on membrane fouling in specific areas in recent years.

Research Area	References
FO; RO; Driven membrane processes; Biofilm dynamics; Membrane performance; Concentration polarization	[2,12,26,56,105]
EPS; SMP; Microbial community structure; Microbial flocs; Microbial soluble substances; Membrane modification	[7,8,15,34,63]
Membrane cleaning; Membrane fouling control; Cross-flow membrane filtration; osmotic pressure	[2,4,10,11,13]
Inherent properties of membrane; Operating conditions; Mixed liquid properties; Fouling mechanisms	[14,22,34,81,88]
Anaerobic membrane bioreactor; Influencing factors; Domestic wastewater; Biosolids production; Energy; Reuse	[23,30,43,45,47,166]
Chemical oxygen demand; SRT; HRT	[22,24,37,60]
Ultrasonication; Hollow fiber membrane; Mathematical model; Emerging micropollutants	[27,44,60,96]
Nutrient recovery; Phosphate recovery; Ammonia recovery; Hybrid system; Direct membrane;	[22,60,112,116,128]

References

1. Li, C.; Deng, W.; Gao, C.; Xiang, X.M.; Feng, X.H.; Batchelor, B.; Li, Y. Membrane distillation coupled with a novel two-stage pretreatment process for petrochemical wastewater treatment and reuse. *Sep. Purif. Technol.* **2019**, *224*, 23–32. [CrossRef]
2. Abdelrasoul, A.; Doan, H.; Lohi, A. Fouling in forward osmosis membranes: Mechanisms, control, and challenges. In *Osmotically Driven Membrane Processes: Approach, Development and Current Status*; IntechOpen Limited: London, UK, 2018; pp. 151–177.
3. Gong, H.; Jin, Z.; Wang, Q.; Zuo, J.; Wu, J.; Wang, K.J. Effects of adsorbent cake layer on membrane fouling during hybrid coagulation/adsorption microfiltration for wastewater organic recovery. *Chem. Eng. J.* **2017**, *317*, 751–757. [CrossRef]
4. Choudhury, M.R.; Anwar, N.; Jassby, D.; Rahaman, M.S. Fouling and wetting in the membrane distillation driven wastewater reclamation process—A review. *Adv. Colloid Interf. Sci.* **2019**, *269*, 370–399. [CrossRef] [PubMed]
5. Sun, S.Y.; Li, X.F.; Feng, L. Characteristics of Extracellular Polysaccharide in Bio-fouling Layer of MBR. *Environ. Sci. Technol.* **2008**, *31*, 99–102.
6. Zhang, K.; Choi, H.; Dionysiou, D.D.; Sorial, G.A.; Oerther, D.B. Identifying pioneer bacterial species responsible for biofouling membrane bioreactors. *Environ. Microbiol.* **2006**, *8*, 433–440. [CrossRef]
7. Wu, B.; Fane, A.G. Microbial relevant fouling in membrane bioreactors: Influencing factors, characterization, and fouling control. *Membranes* **2012**, *2*, 565–584. [CrossRef]
8. Meng, F.G.; Zhang, S.Q.; Oh, Y.; Zhou, Z.B.; Shin, H.S.; Chae, S.R. Fouling in membrane bioreactors: An updated review. *Water Res.* **2017**, *114*, 151–180. [CrossRef]
9. Chen, Y.Q.; Li, F.; Qiao, T.J. Research on ultrafiltration membrane fouling based on chemical cleaning. *China Water Wastewater* **2013**, *29*, 51–54.
10. Blandin, G.; Verliefde, A.R.D.; Comas, J. Efficiently combining water reuse and desalination through forward osmosis-reverse osmosis (FO-RO) hybrids: A critical review. *Membranes* **2016**, *6*, 37. [CrossRef]
11. Bhattacharjee, S.; Kim, A.S.; Elimelech, M. Concentration polarization of interacting solute particles in cross-flow membrane filtration. *Colloid Interface Sci.* **1999**, *212*, 81–99. [CrossRef]
12. Li, C.; Yang, Y.; Ding, S.Y. Dynamics of bio-fouling development on the conditioned membrane and its relationship with membrane performance. *J. Membr. Sci.* **2016**, *514*, 264–273. [CrossRef]

13. Zhou, X.L.; Chen, J.R.; Yu, G.Y.; Hong, H.C.; Jin, L.; Lu, X.F.; Lin, H. Review on mechanism and control of membrane fouling in membrane bioreactor. *Environ. Sci. Technol.* **2012**, *35*, 86–91.
14. Le Clech, P.; Chen, V.; Fane, T.A.G. Fouling in membrane bioreactors used in wastewater treatment. *J. Membr. Sci.* **2006**, *284*, 17–53. [CrossRef]
15. Meng, F.; Chae, S.R.; Drews, A.; Kraume, M.; Shin, H.S.; Yang, F.L. Recent advances in membrane bioreactors (MBRs): Membrane fouling and membrane material. *Water Res.* **2009**, *43*, 1489–1512. [CrossRef] [PubMed]
16. Wang, X.M.; Li, X.Y. Accumulation of biopolymer clusters in a submerged membrane bioreactor and its effect on membrane fouling. *Water Res.* **2008**, *42*, 855–862. [CrossRef] [PubMed]
17. Lin, H.J.; Xie, K.; Mahendran, B. Sludge properties and their effects on membrane fouling in submerged anaerobic membrane bioreactors (SAnMBRs). *Water Res.* **2009**, *43*, 3827–3837. [CrossRef]
18. Wang, L.; Liu, H.B.; Zhang, W.D.; Yu, T.T.; Jin, Q.; Fu, B.; Liu, H. Recovery of organic matters in wastewater by self-forming dynamic membrane bioreactor: Performance and membrane fouling. *Chemosphere* **2018**, *203*, 123–131. [CrossRef]
19. Gao, W.J.; Lin, H.J.; Leung, K.T.; Liao, B.Q. Influence of elevated pH shocks on the performance of a submerged anaerobic membrane bioreactor. *Process Biochem.* **2010**, *45*, 1279–1287. [CrossRef]
20. Arena, J.T.; McCloskey, B.; Freeman, B.D.; McCutcheon, J.R. Surface modification of thin film composite membrane support layers with polydopamine: Enabling use of reverse osmosis membranes in pressure retarded osmosis. *J. Membr. Sci.* **2011**, *375*, 55–62. [CrossRef]
21. Li, X.H. Study of Modified PVDF UF Membrane Anti-Fouling Performance. Master's Thesis, Harbin Institute of Technology, Harbin, China, 2010.
22. Xie, Y.H.; Zhu, T.; Xu, C.H. Research progress in influence factors on membrane fouling in membrane bioreactor. *Chem Eng. (CHINA)* **2010**, *38*, 26–31.
23. Zheng, Y.L.; Li, H.Q.; Liu, L. Research progress in influence factors and control technologies of membrane fouling in anaerobic membrane bioreactor. *Environ. Sci. Technol.* **2015**, *28*, 71–75.
24. Jegatheesan, V.; Pramanik, B.K.; Chen, J.; Navaratna, D.; Chang, C.Y.; Shu, L. Treatment of textile wastewater with membrane bioreactor: A critical review. *Bioresour. Technol.* **2016**, *204*, 202–212. [CrossRef]
25. Yang, F.L.; Zhang, S.T.; Zhang, X.W.; Qu, Y.B.; Liu, Y.H. Experimental study of domestic wastewater treatment with a metal membrane bioreactor. *Desalination* **2005**, *177*, 83–93.
26. Xiao, Q.Q.; Xu, S.C.; Wang, Y. Research and analysis on influencing factors of forward osmosis membrane fouling. *Chem. Ind. Eng. Prog.* **2018**, *37*, 59–367.
27. Mu, S.T.; Fan, H.J.; Han, B.J. Research progress on the process and mathematical model of hollow fiber membrane fouling. *Membr. Sci. Technol.* **2018**, *38*, 114–121.
28. Qu, F.S.; Liang, H.; Wang, Z.Z.; Wang, H.; Yu, H.; Li, G. Ultrafiltration membrane fouling by extracellular organic matters(EOM) of microcystis aeruginosa in stationary phase: Influences of interfacial characteristics of foulants and fouling mechanisms. *Water Res.* **2012**, *46*, 1490–1500. [CrossRef]
29. Katsoufidou, K.S.; Sioutopoulos, D.C.; Yiantsios, S.G.; Karabelas, A.J. UF membrane fouling by mixtures of humic acids and sodium alginate: Fouling mechanisms and reversibility. *Desalination* **2010**, *64*, 220–227. [CrossRef]
30. Shin, C.; Jaeho, B. Current status of the pilot-scale anaerobic membrane bioreactor treatments of domestic wastewaters: A critical review. *Bioresour. Technol.* **2018**, *247*, 1038–1046. [CrossRef]
31. Tian, J.Y.; Ernst, M.; Cui, F.Y.; Jekel, M. Effect of particle size and concentration on the synergistic UF membrane fouling by particles and NOM fractions. *J. Membr. Sci.* **2013**, *446*, 1–9. [CrossRef]
32. Blandin, G.; Gautier, C.; Toran, S.M.; Monclus, H.; Rodriguez-Roda, I.; Comas, J. Retrofitting membrane bioreactor (MBR) into osmotic membrane bioreactor (OMBR): A pilot scale study. *Chem. Eng. J.* **2019**, *339*, 268–277. [CrossRef]
33. Oliver, T.I.; Rania, A.H.; Joo, H.T. Membrane fouling control in membrane bioreactors (MBRs) using granular materials. *Bioresour. Technol.* **2017**, *240*, 9–24.
34. Lin, H.J.; Zhang, M.J.; Wang, F.Y.; Meng, F.; Liao, B.-Q.; Hong, H.; Chen, J.; Gao, W. A critical review of extracellular polymeric substances (EPSs) in membrane bioreactors: Characteristics, roles in membrane fouling and control strategies. *J. Membr. Sci.* **2014**, *460*, 110–125. [CrossRef]
35. Zhou, J.E.; Wang, Y.Q.; Zhang, X.Z. Electrokinetic properties of zirconia/cordierite microfiltration membrane and its influence on permeate flux. *J. Chin. Ceram. Soc.* **2009**, *37*, 299–303.

36. Wu, N.P.; Kong, X.Y.; Fang, S. Study on the application of MBR technology in the recycling of micro-polluted surface water. *Memb. Sci. Technol.* **2016**, *36*, 103–108.
37. Noor, S.A.M.; Zainura, Z.N.; Mohd, A.A.H.; Gustaf, O. Application of membrane bioreactor technology in treating high strength industrial wastewater: A performance review. *Desalination* **2012**, *305*, 1–11.
38. Qiu, H.Y.; Xiao, T.H.; Hu, N.E. The microporous membrane with different pore sizes was used to study the treatment of micro-polluted water in MBR. *Technol. Water Treat.* **2018**, *44*, 94–98.
39. Aslam, M.; Lee, P.H.; Kim, J. Analysis of membrane fouling with porous membrane filters by microbial suspensions for autotrophic nitrogen transformations. *Sep. Purif. Technol.* **2015**, *146*, 284–293. [CrossRef]
40. Kang, S.; Hoekem, V. Effect of membrane surface properties during the fast evaluation of cell attachment. *Sep. Sci. Technol.* **2006**, *41*, 1475–1487. [CrossRef]
41. Fang, H.P.; Shi, X.L. Pore fouling of microfiltration membranes by activated sludge. *Membr. Sci.* **2005**, *264*, 161–166. [CrossRef]
42. Choijh, Y. Effect of membrane type and material on performance of a submerged membrane bioreactor. *Chemosphere* **2008**, *71*, 853–859.
43. Hale, O.; Recep, K.D.; Mustafa, E.E.; Cumali, K.; Spanjers, H.; Lier, J.B.V. A review of anaerobic membrane bioreactors for municipal wastewater treatment: Integration options, limitations and expectations. *Sep. Purif. Technol.* **2013**, *118*, 89–104.
44. Pawel, K.; Lance, L.; Simos, M.; Evina, K. Membrane bioreactors -a review on recent developments in energy reduction, fouling control, novel configurations, LCA and market prospects. *J. Membr. Sci.* **2017**, *527*, 207–227.
45. Lin, H.J.; Peng, W.; Zhang, M.J.; Chen, J.R.; Hong, H.C.; Zhang, Y. A review on anaerobic membrane bioreactors: Applications, membrane fouling and future perspectives. *Desalination* **2013**, *314*, 169–188. [CrossRef]
46. Arabi, S.; Nakhla, G. Impact of protein/carbohydrate ratio in the feed wastewater on the membrane fouling in membrane bioreactors. *J. Membr. Sci.* **2008**, *324*, 142–150. [CrossRef]
47. Chen, J.R.; Zhang, M.J.; Li, F.Q.; Qian, L.; Lin, H.G.; Yang, L.N.; Wu, X.L.; Zhou, X.L.; He, Y.M.; Liao, B.Q. Membrane fouling in a membrane bioreactor: High filtration resistance of gel layer and its underlying mechanism. *Water Res.* **2016**, *102*, 82–89. [CrossRef] [PubMed]
48. Deng, L.J.; Guo, W.S.; Huu, H.N.; Zhang, H.W.; Wang, J.; Li, J.X.; Xia, S.Q.; Wu, Y. Biofouling and control approaches in membrane bioreactors. *Bioresour. Technol.* **2016**, *221*, 656–665. [CrossRef]
49. Vanysacker, L.; Declerck, P.; Bilad, M.R. Biofouling on microfiltration membranes in MBRs: Role of membrane type and microbial community. *J. Membr. Sci.* **2014**, *453*, 394–401. [CrossRef]
50. Zhang, H.F.; Sun, B.S.; Zhao, X.H. Effects of soluble microbial product on the performance of submerged membrane bioreactor. *Environ. Sci.* **2008**, *29*, 82–86.
51. Dukwoo, J.; Youngo, K.; Saeedreza, H.; Yoo, K.; Hoek, E.M.V.; Jeonghwan, K. Biologically induced mineralization in anaerobic membrane bioreactors: Assessment of membrane scaling mechanisms in a long-term pilot study. *J. Membr. Sci.* **2017**, *543*, 342–350.
52. Ji, L.; Zhou, J.T.; Zhang, X.H. Influence of influent composition on membrane fouling in membrane bioreactors. *Environ. Sci.* **2007**, *28*, 18–23.
53. Hwang, B.K.; Lee, W.N.; Yeon, K.M.; Park, C.H.; Chang, I.S.; Drews, A.; Kraume, M. Correlating TMP increases with microbial characteristics in the bio-cake on the membrane surface in a membrane bioreactor. *Environ. Sci. Technol.* **2008**, *42*, 3963–3968. [CrossRef]
54. Lin, H.J.; Lu, X.F.; Duan, W. Filtration characteristics and mechanism of membrane fouling in a membrane bioreactor for municipal wastewater treatment. *Environ. Sci.* **2006**, *27*, 2511–2517.
55. Deng, L.J.; Guo, W.S.; Huu, H.N.; Du, B.; Wei, Q.; Ngoc, H.T.; Nguyen, C.N.; Chen, S.S.; Li, J.X. Effects of hydraulic retention time and bioflocculant addition on membrane fouling in a sponge-submerged membrane bioreactor. *Bioresour. Technol.* **2016**, *210*, 11–17. [CrossRef]
56. Yu, Q.H.; Chi, L.N.; Zhou, W.L. Overview of forward osmosis membrane separation technology: Research and its application to water treatment. *Environ. Sci. Technol.* **2010**, *33*, 117–122.
57. Park, S.; Yeon, K.M.; Moon, S.; Kim, J.O. Enhancement of operating flux in a membrane bio-reactor coupled with a mechanical sieve unit. *Chemosphere* **2018**, *191*, 573–579. [CrossRef] [PubMed]
58. Zhang, C.Y.; Wang, Y.; Huang, X. Experimental study on economic aeration of integrated membrane bioreactor. *Membr. Sci. Technol.* **2004**, *24*, 11–15.

59. Huang, B.C.; Guan, Y.F.; Chen, W.; Yu, H.Q. Membrane fouling characteristics and mitigation in a coagulation–assisted microfiltration process for municipal wastewater pretreatment. *Water Res.* **2017**, *123*, 216–223. [CrossRef]
60. Besha, A.T.; Gebreyohannes, A.Y.; Ashu, T.R. Removal of emerging micropollutants by activated sludge process and membrane bioreactors and the effects of micropollutants on membrane fouling: A review. *J. Environ. Chem. Eng.* **2017**, *5*, 2395–2414. [CrossRef]
61. Wu, H.Y. Study on the Mechanism of Microbial Products in MBR on Membrane Fouling. Master's Thesis, Chinese Research Academy of Environmental Sciences, Beijing, China, 2012.
62. Chin, H.N.; Zainura, Z.N.; Noor, S.A.M. Green technology in wastewater treatment technologies: Integration of membrane bioreactor with various wastewater treatment systems. *Chem. Eng. J.* **2016**, *283*, 582–594.
63. Long, X.Y.; Long, T.R.; Tang, R. SRT on components and surface characters of extracellula polymeric substances. *China Water Wastewater* **2008**, *24*, 1–6.
64. Zhao, J.; Xu, G.T.; Qin, Z. Composing of extracellular polymeric substances and its effect on sludge characteristics. *Safety Environ. Eng.* **2008**, *15*, 66–69.
65. Tao, L.J.; Li, X.F.; Wang, X.H. Effects of sludge retention time on sludge characteristics in membrane bioreactors. *Chinese J. Environ. Eng.* **2012**, *6*, 719–724.
66. Recep, K.D.; Aurelie, G.; Barry, H.; Frank, P.Z.; Jules, B.L. Implications of changes in solids retention time on long term evolution of sludge filterability in anaerobic membrane bioreactors treating high strength industrial wastewater. *Water Res.* **2014**, *59*, 11–22.
67. Tian, Y.; Li, H.; Li, L.P. In-situ integration of microbial fuel cell with hollow-fiber membrane bioreactor for wastewater treatment and membrane fouling mitigation. *Biosens. Bioelectron.* **2015**, *64*, 189–195. [CrossRef] [PubMed]
68. Alturki, A.; McDonald, J.; Khan, S.J.; Hai, F.L.; Price, W.E.; Nghiem, L.D. Performance of a novel osmotic membrane bioreactor (OMBR) system: Flux stability and removal of trace organics. *Bioresour. Technol.* **2012**, *113*, 201–206. [CrossRef] [PubMed]
69. Aslam, M.; McCarty, P.L.; Shin, C.; Bae, J.; Kim, J. Low energy single-staged anaerobic fluidized bed ceramic membrane bioreactor (AFCMBR) for wastewater treatment. *Bioresour. Technol.* **2017**, *240*, 33–41. [CrossRef]
70. Jin, Z.; Meng, F.; Gong, H.; Wang, C.; Wang, K. Improved low-carbon-consuming fouling control in long-term membrane- based wastewater pre-concentration: The role of enhanced coagulation process and air backflushing in sustainable wastewater treatment. *Membr. Sci.* **2017**, *529*, 252–262. [CrossRef]
71. Ali, S.M.; Kim, J.E.; Phuntsho, S.; Jang, A.; Choi, J.Y.; Shon, H.K. Forward osmosis system analysis for optimum design and operating conditions. *Water Res.* **2018**, *145*, 429–441. [CrossRef]
72. Kim, S.R.; Lee, K.B.; Kim, J.E.; Won, Y.J.; Yeon, K.M. Macroencapsulation of quorum quenching bacteria by polymeric membrane layer and its application to MBR for biofouling control. *J. Membr. Sci.* **2015**, *473*, 109–117. [CrossRef]
73. Hai, F.I.; Riley, T.; Shawkat, S.; Magram, S.F.; Yamamoto, K. Removal of pathogens by membrane bioreactors: A review of the mechanisms, influencing factors and reduction in chemical disinfectant dosing. *Water* **2014**, *6*, 3603–3630. [CrossRef]
74. Arevalo, J.; Ruiz, L.M.; P erez, J. Effect of temperature on membrane bioreactor performance working with high hydraulic and sludge retention time. *Biochem. Eng. J.* **2014**, *88*, 42–49. [CrossRef]
75. Charfi, A.; Aslam, M.; Lesage, G.; Heran, M.; Kim, J. Macroscopic approach to develop fouling model under GAC fluidization in anaerobic fluidized bed membrane bioreactor. *J. Ind. Eng. Chem.* **2017**, *49*, 219–229. [CrossRef]
76. Liu, Y.L.; Xie, Y.M.; Wang, X.X. fouling analysis and chemical cleaning of reverse osmosis membrane in waste water reuse project of steel plant. *Clean. World* **2019**, *35*, 8–10.
77. Achilli, A.; Cath, T.Y.; Marchand, E.A.; Childress, A.E. The forward osmosis membrane bioreactor: A low fouling alternative to MBR processes. *Desalination* **2009**, *239*, 10–21. [CrossRef]
78. Wang, Z.W.; Wu, Z.C.; Gu, G.W. Study on operation mode of SMBR process for treatment of wastewater. *Environ. Eng.* **2005**, *23*, 7–9.
79. Xie, Y.H.; Zhu, T.; Xu, C.H. Membrane cleaning method in metal membrane bioreactor. *Chem. Eng.* **2010**, *38*, 190–193.
80. Hong, Y.; Xiao, P.; Dong, W. Membrane fouling and chemical cleaning for wastewater reclamation using submerged ultrafiltration membrane. *Chin. J. Environ. Eng.* **2016**, *10*, 2495–2500.

81. Guo, W.S.; Ngo, H.H.; Li, J.X. A mini-review on membrane fouling. *Bioresour. Technol.* **2017**, *122*, 27–34. [CrossRef]
82. Wang, Z.W.; Wu, Z.C. A review of membrane fouling in MBRs: Characteristics and role of sludge cake formed on membrane surfaces. *Sep. Sci. Technol.* **2009**, *44*, 3571–3596. [CrossRef]
83. Bell, E.A.; Holloway, R.W.; Cath, T.Y. Evaluation of forward osmosis membrane performance and fouling during long-term osmotic membrane bioreactor study. *J. Membr. Sci.* **2016**, *517*, 1–13. [CrossRef]
84. Chen, C.H.; Fu, Y.; Gao, D.W. Membrane biofouling process correlated to the microbial community succession in an A/O MBR. *Bioresour. Technol.* **2015**, *197*, 185–192. [CrossRef]
85. Abdessemed, D. Treatment of primary effluent by coagulation-adsorption-ultrafiltration for reuse. *Desalination* **2002**, *152*, 367–373. [CrossRef]
86. Zinadini, S.; Vatanpour, V.; Zinatizadeh, A.A.; Rahimi, M.; Rahimi, Z.; Kian, M. Preparation and characterization of antifouling graphene oxide/polyethersulfone ultrafiltration membrane: Application in MBR for dairy wastewater treatment. *J. Water Process Eng.* **2015**, *7*, 280–294. [CrossRef]
87. Zhang, J.L.; Cao, Z.P.; Zhang, H.W. Effects of sludge retention time (SRT) on the characteristics of membrane bioreactor (MBR). *Environ. Sci.* **2008**, *29*, 2788–2793.
88. Wang, Z.W.; Ma, J.X.; Tang, C.Y.Y.; Kimura, K.; Wang, Q.Y.; Han, X.M. Membrane cleaning in membrane bioreactors: A review. *J. Membr. Sci.* **2014**, *468*, 276–307. [CrossRef]
89. Fu, C.; Yue, X.D.; Shi, X.Q.; Ng, K.K.; Ng, H.Y. Membrane fouling between a membrane bioreactor and a moving bed membrane bioreactor: Effects of solids retention time. *Chem. Eng. J.* **2017**, *309*, 397–408. [CrossRef]
90. Li, S.F.; Cui, C.W.; Huang, J. Effect of extracellular polymeric substances on membrane fouling of membrane bioreactor. *J. Harbin Inst. Technol.* **2007**, *39*, 266–269.
91. Shen, L.G.; Lei, Q.; Chen, J.R.; Hong, H.C.; He, Y.M.; Lin, H.J. Membrane fouling in a submerged membrane bioreactor: Impacts of floc size. *Chem. Eng. J.* **2015**, *269*, 328–334. [CrossRef]
92. Hu, Y.S.; Wang, X.C.; Yu, Z.Z.; Ngo, H.H.; Sun, Q.Y.; Zhang, Q.H. New insight into fouling behavior and foulants accumulation property of cake sludge in a full-scale membrane bioreactor. *J. Membr. Sci.* **2016**, *510*, 10–17. [CrossRef]
93. Wang, L.L.; Song, W.C. Study on chemical cleaning technology of membrane bioreactor. *Water Sci. Eng. Technol.* **2018**, *1*, 68–70.
94. Oh, K.S.; Poh, P.E.; Chong, M.N.; Chan, E.S.; Lau, E.V.; Saint, C.P. Bathroom greywater recycling using polyelectrolyte-complex bilayer membrane: Advanced study of membrane structure and treatment efficiency. *Carbohydr. Polym.* **2016**, *148*, 161–170. [CrossRef] [PubMed]
95. Li, S.F.; Gao, Y. Effect of powdered activated carbon on the sludge mixed liquor characteristics and membrane fouling of MBR. *Environ. Sci.* **2012**, *32*, 508–514.
96. Samira, A.O.; Alireza, K.; Mahdie, S.; Yasin, O.; Vahid, V. A review on the applications of ultrasonic technology in membrane bioreactors. *Ultrason. Sonochem.* **2019**, *58*, 104633.
97. Laqbaqbi, M.; García-Payo, M.C.; Khayet, M.; EI Kharraz, J.; Chaouch, M. Application of direct contact membrane distillation for textile wastewater treatment and fouling study. *Sep. Purif. Technol.* **2019**, *209*, 815–825. [CrossRef]
98. Tian, Y.; Li, Z.N.; Chen, L. Fouling property and interaction energy with EPS membrane in normal sludge and bulking sludge. *Environ. Sci.* **2013**, *33*, 1224–1230.
99. Berkessa, Y.W.; Yan, B.H.; Li, T.F.; Tan, M.; She, Z.L.; Jegatheesan, V.; Jiang, H.Q.; Zhang, Y. Novel anaerobic membrane bioreactor (AnMBR) design for wastewater treatment at long HRT and high solid concentration. *Bioresour. Technol.* **2018**, *250*, 281–289. [CrossRef]
100. Chen, X.G.; Li, G.; Lin, H.B. Operation performance and membrane fouling of a spiral symmetry stream anaerobic membrane bioreactor supplemented with biogas aeration. *J. Membr. Sci.* **2017**, *539*, 206–212.
101. Wang, Z.Z.; Meng, F.G.; He, X.; Zhou, Z.B.; Huang, L.N.; Liang, S. Optimisation and performance of NaClO-assisted maintenance cleaning for fouling control in membrane bioreactors. *Water Res.* **2014**, *53*, 1–11. [CrossRef]
102. Amine, C.; Eunyoung, P.; Muhammad, A.; Kim, J. Particle-sparged anaerobic membrane bioreactor with fluidized polyethylene terephthalate beads for domestic wastewater treatment: Modelling approach and fouling control. *Bioresour. Technol.* **2018**, *258*, 263–269.

103. Zhou, Z.B.; Meng, F.G.; Lu, H.; Li, Y.; Jia, X.S.; He, X. Simultaneous alkali supplementation and fouling mitigation in membrane bioreactors by on-line NaOH backwashing. *J. Membr. Sci.* **2014**, *457*, 120–127. [CrossRef]
104. Judd, S.J. The status of industrial and municipal effluent treatment with membrane bioreactor technology. *Chem. Eng. J.* **2016**, *305*, 37–45. [CrossRef]
105. Ly, Q.V.; Hu, Y.X.; Li, J.X.; Cho, J.; Hur, J. Characteristics and influencing factors of organic fouling in forward osmosis operation for wastewater applications: A comprehensive review. *Environ. Int.* **2019**, *129*, 164–184. [CrossRef] [PubMed]
106. Mezohegyi, G.; Bilad, M.R.; Vankelecom, I.F.J. Direct wastewater up-concentration by submerged aerated and vibrated membranes. *Bioresour. Technol.* **2012**, *118*, 1–7. [CrossRef] [PubMed]
107. Naidu, G.; Jeong, S.; Choi, Y.; Vigneswaran, S. Membrane distillation for wastewater reverse osmosis concentrate treatment with water reuse potential. *J. Membr. Sci.* **2017**, *524*, 565–575. [CrossRef]
108. Xiong, J.Q.; Yu, S.C.; Hu, Y.S.; Yang, Y.; Wang, X.C.C. Applying a dynamic membrane filtration (DMF) process for domestic wastewater preconcentration: Organics recovery and bioenergy production potential analysis. *Sci. Total Environ.* **2019**, *68*, 35–43. [CrossRef]
109. Meng, F.G.; Zhang, H.M.; Yang, F.L. Effectof filamentous bacteriaon membrane foulingin submerged membrane bioreactor. *J. Membr Sci.* **2006**, *272*, 161–168. [CrossRef]
110. Zhang, Y.; Zhao, P.; Li, J.; Hou, D.Y.; Wang, J.; Liu, H.J. A hybrid process combining homogeneous catalytic ozonation and membrane distillation for wastewater treatment. *Chemosphere* **2016**, *160*, 134–140. [CrossRef]
111. Wang, L. Gas/liquid membrane contact separation process and its membrane material research. *Light Ind. Sci. Technol.* **2013**, *4*, 46–60.
112. Yan, T.; Ye, Y.Y.; Ma, H.M. A critical review on membrane hybrid system for nutrient recovery from wastewater. *Chem. Eng. J.* **2018**, *348*, 143–156. [CrossRef]
113. Wang, Z.; Lv, Y.W.; Wang, S.M. Preparation and characterization of PVDF/CA blend ultrafiltration membrane. *Membr. Sci. Technol.* **2002**, *22*, 4–8.
114. Sun, Q. Preparation and antifouling property of polyacrylonitrile (PAN)-sulfobetaine copolymer ultrafiltration membrane. Master's Thesis, Tianjin University, Tian Jin, China, 2007.
115. Li, B.R.; Meng, M.J.; Cui, Y.H.; Wu, Y.L.; Zhang, Y.L.; Dong, H.J.; Zhu, Z.; Feng, Y.H.; Wu, C.D. Changing conventional blending photocatalytic membranes (BPMs): Focus on improving photocatalytic performance of $Fe_3O_4/g-C_3N_4$/PVDF membranes through magnetically induced freezing casting method. *Chem. Eng. J.* **2019**, *365*, 405–414. [CrossRef]
116. Selina, H.; Majid, E. Direct membrane filtration for wastewater treatment and resource recovery: A review. *Sci. Total Environ.* **2020**, *710*, 136375.
117. Ng, D.Y.F.; Wu, B.; Chen, Y.; Dong, Z.; Wang, R. A novel thin film composite hollow fiber osmotic membrane with one-step prepared dual-layer substrate for sludge thickening. *J. Membr. Sci.* **2019**, *575*, 98–108. [CrossRef]
118. Xu, J.; Tran, T.N.; Lin, H.; Dai, N. Removal of disinfection byproducts in forward osmosis for wastewater recycling. *J. Membr. Sci.* **2018**, *564*, 352–360. [CrossRef]
119. Ye, Z.L.; Ghyselbrecht, K.; Monballiu, A.; Pinoy, L.; Meesschaert, B. Fractionating various nutrient ions for resource recovery from swine wastewater using simultaneous anionic and cationic selective-electrodialysis. *Water Res.* **2019**, *160*, 424–434. [CrossRef]
120. Meng, F.; Chae, S.R.; Shin, H.S. Recent advances in membrane bioreactors: configuration development, pollutant elimination, and sludge reduction. *Environ. Eng. Sci.* **2011**, *29*, 139–160. [CrossRef]
121. Zuthi, M.F.R.; Ngo, H.H.; Guo, W.S. A review towards finding a simplified approach for modelling the kinetics of the soluble microbial products (SMP) in an integrated mathematical model of membrane bioreactor (MBR). *Int. Biodeterior. Biodegrad.* **2013**, *85*, 466–473. [CrossRef]
122. Yoon, Y.; Hwang, Y.; Kwon, M.; Jung, Y.; Hwang, T.M.; Kang, J.W. Application of O-3 and O-3/H_2O_2 as post-treatment processes for color removal in swine wastewater from a membrane filtration system. *J. Ind. Eng. Chem.* **2014**, *20*, 2801–2805. [CrossRef]
123. Zarebska, A.; Nieto, D.R.; Christensen, K.V.; Norddahl, B. Ammonia recovery from agricultural wastes by membrane distillation: Fouling characterization and mechanism. *Water Res.* **2014**, *56*, 1–10. [CrossRef]
124. Wu, W.Z.; Chen, G.E. Research progress in modification of PVDF membranes. *J. Shanghai Inst. Technol. (Nat. Sci.)* **2013**, *13*, 118–127.

125. Bagastyo, A.Y.; Anggrainy, A.D.; Nindita, C.S. Electrodialytic removal of fluoride and calcium ions to recover phosphate from fertilizer industry wastewater. *Sustain. Environ. Res.* **2017**, *27*, 230–237. [CrossRef]
126. Zhao, X.; Xuan, H.; Chen, Y.; He, C. Preparation and characterization of superior antifouling PVDF membrane with extremely ordered and hydrophilic surface layer. *J. Membr. Sci.* **2015**, *494*, 48–56. [CrossRef]
127. Zhang, Y.C.; Yang, Q.; Chang, Q. Low temperature plasma modification and electrokinetic characteristics of PVDF hollow fiber membrane. *Technol. Water Treat.* **2011**, *37*, 57–60.
128. Zhan, W.; Zhang, J.; Zuo, W. Research progress in hydrophilic and photocatalytic modification of organic polymer membrane. *Environ. Sci. Technol.* **2017**, *40*, 114–119.
129. Wang, L.; Wei, J.F.; Chen, Y. Preparation of novel hydrophilic PVDF hollow fiber membrane. *J. Tianjin Polytech. Univ.* **2013**, *32*, 7–10.
130. Cui, K.L.; Meng, S.Q.; Gao, B.J. Adsorption and separation of toxic anions on polysulfone microfiltration membrane surface grafted with tertiary amine monomer DMAEMA and grafted membrane. *Polym. Mater. Sci. Eng.* **2017**, *33*, 137–143.
131. Wang, A.; Wei, J.F.; Zhao, K.Y. Study on UV Grafting Modification of Polypropylene Film. *J. Funct. Mater.* **2012**, *43*, 2851–2854.
132. Zhao, Y.H.; Wee, K.H.; Bai, R. Highly hydrophilic and low-protein-fouling polypropylene membrane prepared by surface modification with sulfobetaine-based zwitterionic polymer through a combined surface polymerization method. *J. Membr. Sci.* **2010**, *362*, 326–333. [CrossRef]
133. Jin, G.; Wei, J.F.; Wang, A. Study of grafting sodium p-styrene sulfonate onto surface of polypropylene hollow fiber membrane. *Mod. Chem. Ind.* **2018**, *38*, 98–101.
134. Wang, L.; Wei, J.F.; Zhao, K.Y. Preparation and characterization of high-hydrophilic polyhydroxy functional PP hollow fiber membrane. *Mater. Lett.* **2015**, *159*, 189–192. [CrossRef]
135. Hu, M.X.; Yang, Q.; Xu, Z.K. Enhancing the hydrophilicity of polypropylene microporous membranes by the grafting of 2-hydroxyethyl methacrylate via a synergistic effect of photoinitiators. *J. Membr. Sci.* **2006**, *285*, 196–205. [CrossRef]
136. Xiong, C.C.; Li, W.X.; Liu, Y.F. Optimization membrane module structure of column type by CFD. *CIESC J.* **2017**, *68*, 4341–4350.
137. Zavala, M.Á.L.; Pérez, L.B.S.; Reynoso-Cuevas, L.; Funamizu, N. Pre-filtration for enhancing direct membrane filtration of graywater from washing machine discharges. *Ecol. Eng.* **2014**, *64*, 116–119. [CrossRef]
138. Xu, X.J. New spiral membrane module and its application in water treatment. Master's Thesis, Dalian University of Technology, Dalian, China, 2010.
139. Benvenuti, T.; Krapf, R.S.; Rodrigues, M.A.S. Recovery of nickel and water from nickel electroplating wastewater by electrodialysis. *Sep. Purif. Technol.* **2014**, *129*, 106–112. [CrossRef]
140. Benvenuti, T.; Siqueira Rodrigues, M.A.; Bernardes, A.M.; Zoppas-Ferreira, J. Closing the loop in the electroplating industry by electrodialysis. *J. Clean. Prod.* **2017**, *155*, 130–138. [CrossRef]
141. Merkel, A.; Ashrafi, A.M.; Ondrušek, M. The use of electrodialysis for recovery of sodium hydroxide from the high alkaline solution as a model of mercerization wastewater. *J. Water Process Eng.* **2017**, *20*, 123–129. [CrossRef]
142. Drews, A. Membrane fouling in membrane bioreactors characterisation, contradictions, cause and cures. *J. Membr. Sci.* **2010**, *363*, 1–28. [CrossRef]
143. El-Abbassi, A.; Kiai, H.; Hafidi, A.; Garcia-Payo, M.C.; Khayet, M. Treatment of olive mill wastewater by membrane distillation using polytetrafluoroethylene membranes. *Sep. Purif. Technol.* **2012**, *98*, 55–61. [CrossRef]
144. Zhang, T.; Li, C.X.; Guo, K. Swirling flow enhanced hollow fiber membrane module structure optimization and shell-side hydrodynamics investigation. *CIESC J.* **2018**, *10*, 1–19.
145. Zhang, J.P.; Zhang, H.M.; Yang, F.L. Optimization design of new folding plate module and its application in MBR. *Membr. Sci. Technol.* **2015**, *35*, 72–80.
146. Damtie, M.M.; Kim, B.; Woo, Y.C.; Choi, J.S. Membrane distillation for industrial wastewater treatment: Studying the effects of membrane parameters on the wetting performance. *Chemosphere* **2018**, *206*, 793–801. [CrossRef] [PubMed]
147. El-Abbassi, A.; Hafidi, A.; García-Payo, M.C.; Khayet, M. Concentration of olive mill wastewater by membrane distillation for polyphenols recovery. *Desalination* **2009**, *245*, 670–674. [CrossRef]

148. Li, Y.Y.; Zhao, Y.H.; Yang, J. Influence of adding zeolite on membrane filtration resistance in MBR and its decolorizing effect. *China Water Wastewater* **2008**, *24*, 49–51.
149. Djouadi Belkada, F.; Kitous, O.; Drouiche, N.; Aoudj, S.; Bouchelaghem, O.; Abdi, N.; Grib, H.; Mameri, N. Electrodialysis for fluoride and nitrate removal from synthesized photovoltaic industry wastewater. *Sep. Purif. Technol.* **2018**, *204*, 108–115. [CrossRef]
150. Wu, J.L.; Chen, F.T.; Huang, X. Using inorganic coagulants to control membrane fouling in a submerged membrane bioreactor. *Desalination* **2006**, *197*, 124–136. [CrossRef]
151. Zhang, J.; Padmasiri, S.I.; Fitch, M. Influence of cleaning frequency and membrane history on fouling in an anaerobic membrane bioreactor. *Desalination* **2007**, *207*, 153–166. [CrossRef]
152. Du, X.; Zhang, Z.; Carlson, K.H.; Lee, J.; Tong, T.Z. Membrane fouling and reusability in membrane distillation of shale oil and gas produced water: Effects of membrane surface wettability. *J. Membr. Sci.* **2018**, *567*, 199–208. [CrossRef]
153. Zhang, G.Q.; Li, P.; Qin, X.Q. Ultrasonic control membrane bioreactor membrane fouling. *Low Temp. Build. Technol.* **2011**, *11*, 13–14.
154. El-Abbassi, A.; Hafidi, A.; Khayet, M. Integrated direct contact membrane distillation for olive mill wastewater treatment. *Desalination* **2013**, *323*, 31–38. [CrossRef]
155. Wu, J.L.; Huang, X. Use of ozonation to mitigate fouling in a long-term membrane bioreactor. *Bioresour. Technol.* **2010**, *101*, 6019–6027. [CrossRef]
156. Dai, J.; Deng, Z.J.; Chen, L.W. Research progress on membrane fouling control technology in membrane bioreactor. *Water Treat. Technol.* **2013**, *39*, 10–13.
157. Kim, J.; Kim, K.; Ye, H.; Lee, E.; Shin, C.; McCarty, P.L.; Bae, J. Anaerobic fluidized bed membrane bioreactor for wastewater treatment. *Environ. Sci. Technol.* **2010**, *45*, 576–581. [CrossRef] [PubMed]
158. Hong, S.P.; Bae, T.H.; Tak, T.M. Fouling control in activated sludge submerged hollow fiber membrane bioreactors. *Desalination* **2002**, *143*, 219–228. [CrossRef]
159. Jiang, T.; Kennedy, M.D.; Guinzbourg, B.F.; Vanrolleghem, P.A.; Schippers, J.C. Optimising the operation of a MBR pilot plant by quantitative analysis of the membrane fouling mechanism. *Water. Sci. Treat.* **2005**, *51*, 19–25. [CrossRef]
160. Fan, Y.B.; Wang, J.S.; Jiang, Z.C. Optimal backwash cycle for membranes in membrane bioreactors. *J. Environ. Sci.* **1997**, *17*, 439–444.
161. Wen, X.; Sui, P.; Huang, X. Exerting ultrasound to control the membrane fouling in filtration of anaerobic activated sludge-mechanism and membrane damage. *Water Sci. Technol.* **2008**, *57*, 773–779. [CrossRef] [PubMed]
162. Porcellin, J.S. Chemical cleaning of potable water membranes: A review. *Sep. Purif. Treat.* **2010**, *71*, 137–143. [CrossRef]
163. Viet, N.D.; Cho, J.; Yoon, Y.; Jang, A. Enhancing the removal efficiency of osmotic membrane bioreactors: A comprehensive review of influencing parameters and hybrid configurations. *Chemosphere* **2019**, *236*, 124363. [CrossRef]
164. Gao, Y.; Fang, Z.; Liang, P.; Huang, X. Direct concentration of municipal wastewater by forward osmosis and membrane fouling behavior. *Bioresour. Technol.* **2018**, *247*, 730–735. [CrossRef]
165. Bae, J.; Shin, C.; Lee, E. Anaerobic treatment of lowstrength wastewater: A comparison between single and staged anaerobic fluidized bed membrane bioreactors. *Bioresour. Technol.* **2014**, *165*, 75–80. [CrossRef]
166. Crone, B.C.; Garland, J.L.; Sorial, G.A.; Vane, L.M. Significance of dissolved methane in effluents of anaerobically treated low strength wastewater and potential for recovery as an energy product: A review. *Water Res.* **2016**, *104*, 520–531. [CrossRef] [PubMed]
167. Dong, Q.; Parker, W.; Dagnew, M. Long term performance of membranes in an anaerobic membrane bioreactor treating municipal wastewater. *Chemosphere* **2016**, *144*, 249–256. [CrossRef] [PubMed]
168. Krzeminski, P.; Van der Graaf, J.H.; van Lier, J.B. Specific energy consumption of membrane bioreactor (MBR) for wastewater treatment. *Water Sci. Technol.* **2012**, *65*, 380–392. [CrossRef] [PubMed]
169. Lin, H.; Chen, J.; Wang, F.; Ding, L.; Hong, H. Feasibility evaluation of submerged anaerobic membrane bioreactor for municipal secondary wastewater treatment. *Desalination* **2011**, *280*, 120–126. [CrossRef]
170. Martin, I.; Pidou, M.; Soares, A.; Judd, S.; Jefferson, B. Modelling the energy demands of aerobic and anaerobic membrane bioreactors for wastewater treatment. *Environ. Technol.* **2011**, *32*, 921–932. [CrossRef]

171. Hong, H.; Zhang, M.; He, Y.; Chen, J.R.; Lin, H.J. Fouling mechanisms of gel layer in a submerged membrane bioreactor. *Bioresour. Technol.* **2014**, *166*, 295–302. [CrossRef]
172. Xu, M.; Wen, X.; Huang, X.; Li, Y.S. Membrane fouling control in an anaerobic membrane bioreactor coupled with online ultrasound equipment for digestion of waste activated sludge. *Sep. Sci. Technol.* **2010**, *45*, 941–947. [CrossRef]
173. Sui, P.; Wen, X.; Huang, X. Feasibility of employing ultrasound for on-line membrane fouling control in an anaerobic membrane bioreactor. *Desalination* **2008**, *219*, 203–213. [CrossRef]
174. Ruiz, L.; Perez, J.; Gómez, A.; Letona, A.; Gomez, M.A. Ultrasonic irradiation for ultrafiltration membrane cleaning in MBR systems: Operational conditions and consequences. *Water Sci. Technol.* **2017**, *75*, 802–812. [CrossRef]

© 2020 by the authors. Licensee MDPI, Basel, Switzerland. This article is an open access article distributed under the terms and conditions of the Creative Commons Attribution (CC BY) license (http://creativecommons.org/licenses/by/4.0/).

Review

Recent Advances in the Prediction of Fouling in Membrane Bioreactors

Yaoke Shi [1], Zhiwen Wang [1,2,3], Xianjun Du [1,2,3,4,*], Bin Gong [1], Veeriah Jegatheesan [4] and Izaz Ul Haq [1]

1. Department of Automation, College of Electrical and Information Engineering, Lanzhou University of Technology, Lanzhou 730050, China; yaoke_shi@163.com (Y.S.); wwwangzhiwen@163.com (Z.W.); gong_bin01@163.com (B.G.); Izaz.lut@gmail.com (I.U.H.)
2. Key Laboratory of Gansu Advanced Control for Industrial Processes, Lanzhou University of Technology, Lanzhou 730050, China
3. National Demonstration Center for Experimental Electrical and Control Engineering Education, Lanzhou University of Technology, Lanzhou 730050, China
4. School of Engineering, RMIT University, Melbourne 3000, Australia; jega.jegatheesan@rmit.edu.au
* Correspondence: xdu@lut.edu.cn

Abstract: Compared to the traditional activated sludge process, the membrane bioreactor (MBR) has several advantages such as the production of high-quality effluent, generation of low excess sludge, smaller footprint requirements, and ease of automatic control of processes. The MBR has a broader prospect of its applications in wastewater treatment and reuse. However, membrane fouling is the biggest obstacle for its wider application. This paper reviews the techniques available to predict fouling in MBR, discusses the problems associated with predicting fouling status using artificial neural networks and mathematical models, summarizes the current state of fouling prediction techniques, and looks into the trends in their development.

Keywords: artificial neural network; mathematical model; membrane bioreactor; fouling prediction

1. Introduction

Under the circumstances where water pollution is a great concern and the increasing demand for treatment efficiency and effluent quality are of paramount importance, membrane bioreactors (MBR) [1] have been widely used, based on their merits in addressing those issues. MBR technology is recognized globally as one of the most potential high-tech applications in the field of water treatment in the 21st century. MBR is a wastewater treatment system that combines membrane technology and biological treatment technology [2], and mainly composed of membrane modules and bioreactors [3–5]. It does not need secondary clarification and has the advantages of small area requirement, good effluent quality, and low sludge production. However, it also has the disadvantages of high cost, high energy consumption, and its membranes can be fouled easily [6,7]. Membrane fouling can shorten membrane life and cause unnecessary loss of productivity, which is one of the important reasons limiting the development of MBR for wider applications [8]. Before membrane fouling occurs, timely cleaning or replacement of membrane modules can effectively prolong the service life of the membranes and reduce operating costs [9,10]. Therefore, it is very important to predict membrane fouling and in order to do so factors such as membrane flux, transmembrane pressure (TMP), related operating conditions and predicting their correlation need to be researched [11–14]. The authors reviewed the literature on MBR membrane fouling prediction published worldwide and found that there is scarce literature on the prediction of fouling of membranes in MBRs. Therefore, research outcomes on the prediction of membrane fouling are reviewed in this work, which hopes to be helpful for researchers conducting further study in the future. Figure 1 is an overview of fouling prediction methods.

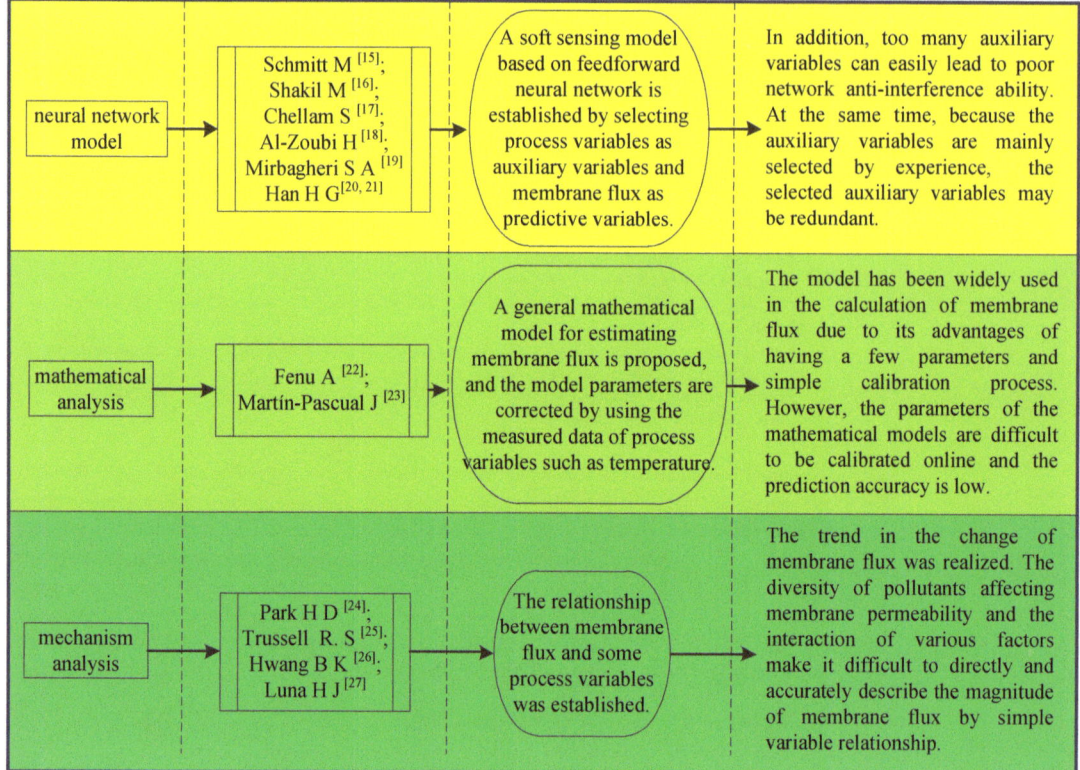

Figure 1. An overview of fouling prediction methods [15–27].

2. Method Based on Artificial Neural Networks to Predict Membrane Fouling

As a mathematical processing method to simulate the structure and function of the biological neural system, an artificial neural network (ANN) has the ability to learn from existing information on the inputs of a process and produce results similar to the ones obtained from the process. It includes three main parts: an input layer of neurons, a hidden layer of neurons, and an output layer of neurons, and ANN is often used to solve problems such as classification and regression [28]. After years of research and exploration, it has been found to have a strong advantage in the field of prediction. The prediction method based on ANN aims to take the original measurement data or the features extracted from the original measurement data as the input of the network, to continuously adjust the structure and parameters of the network through certain training algorithms, and to use the optimized network for prediction. No prior information is needed in the prediction process, and the prediction results are completely based on the monitoring data [29]. For the complexity of membrane fouling prediction, ANN is undoubtedly a simple method, which can connect the input variables in wastewater treatment operation to the membrane fouling status without any mechanistic equations [30]. The simple structure of ANN makes its application and processing simple and efficient. However, when the input data is not prepared well enough or the structure is too complex, the neural network used to predict membrane fouling is easy to be overfitted. In fact, collecting a large number of data increases the processing time and the operating costs [31]. Meanwhile, an over-optimized ANN will also easily lead to overfitting [32]. Therefore, scholars worldwide have carried out much research and tried to find a balance between performance and feasibility. Some studies have focused on the development of ANNs considering pilot-scale MBRs, and achieved good results in predicting the TMP, permeability, and flux of a MBR. To apply ANN to the prediction of membrane fouling in a large-scale membrane bioreactor, it is necessary to carry out in-depth research on this subject. The number of papers published in this area between 2010 and 2021 using Baidu Scholar search engine is shown in Figure 2. The average annual growth rate since 2010 is 22.7%.

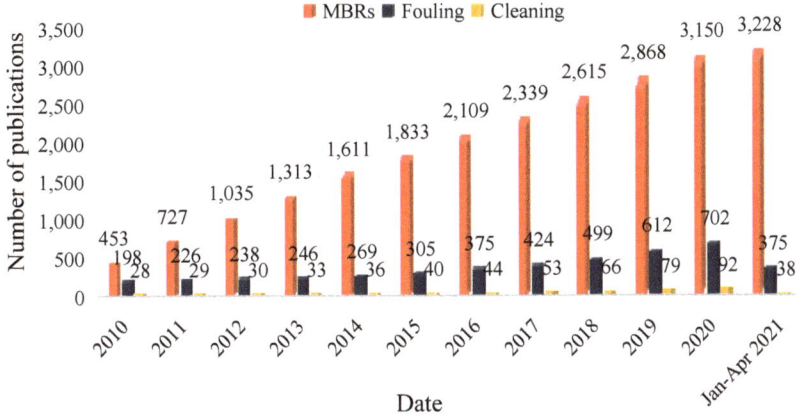

Figure 2. Number of papers published between 2010 and 2021 on MBRs and their fouling and cleaning.

As early as 2000, some scholars began research on ANN-based membrane fouling prediction. Dornier et al. [33] first applied ANN to membrane fouling and successfully predicted the fouling of membrane by sugarcane juice diluent filtered by a ceramic microfiltration membrane under short-term operations. However, the selection of data for network model learning and validation needed further research. Then, back-propagation neural network (BPNN) predicted the membrane fouling of municipal drinking water filtered by nanofiltration membrane, and the network took the total resistance as the prediction output to characterize the status of membrane fouling [34]. The TMP difference can reveal membrane fouling to a great extent while the membrane filtration resistance characterizes it. ANN is used to study the variation trend in the TMP difference before and after backwashing, and a recurrent neural network model for the long-term behavior of membrane filtration was established. After inputting the operating conditions, as long as an initial resistance of membrane filtration is an input, the membrane filtration resistance value in the following days can be predicted accurately, which can be used to predict the status of membrane fouling [35].

Since then, the prediction of membrane fouling in different types of membrane bioreactors has been verified with the development of ANNs. Schmitt et al. [15] used an ANN model to predict TMP difference in the process of an anoxic–aerobic membrane bioreactor (AO–MBR) treating domestic wastewater. Ten parameters related to wastewater treatment measured at different locations of the AO–MBR system were used as input variables of the ANN model, which verified that ANN was an effective way to predict membrane fouling of an AO–MBR system treating domestic wastewater.

Another study used ANNs to predict the flux evolution of MBR on a pilot scale [36]. The wastewater treatment system consisted of two different MBR units. Each unit had three hollow fiber membrane modules. An Elman network was used for modeling, which is composed of a standard feed-forward neural network (FFNN) with a single hidden layer, only adding a feedback connection between the output of the hidden layer and its input. The permeate flux was the output of the ANN model, and the input data include TMP, rate of change of TMP, TMP during the backwash, length of filtration cycle and backwash, sludge retention time (SRT), total suspended solids (TSS), temperature, and oxygen decay rate in the aerobic zone. During the experiment, TMP increased gradually, ranging from 0 to 40 kPa. The average deviation was only 2.7% after modeling the permeate flux. However, the parameters related to wastewater characteristics or mixture were not considered in their work, which leads to a lack of reliability in predicting membrane fouling. Moreover, due to the high number of hidden neurons, the predicted value of permeate flux may exceed the experimental data. The problem of over fitting will appear when there are more hidden layers. At the same time, it will also make the training and the convergence of the model difficult.

In recent years, soft sensing technology has been used widely in the identification and prediction of variables in the wastewater treatment process based on the characters of economic reliability, rapid dynamic response, and high accuracy of identification and prediction. This has become a new approach in membrane flux prediction [16,37]. Chellam [17] and Al-Zoubi et al. [18] obtained multiple process variables related to membrane permeability by analyzing the mechanisms, and 29 of them were selected as auxiliary variables to establish a soft sensing model of membrane permeability based on BPNN. The prediction accuracy reached 70%. However, due to the selection of too many auxiliary variables and the large initial scale of the network, the learning time of the network was very long. Too many auxiliary variables also led to the poor anti-interference ability of the network. Therefore, it can only be used in the pilot platform, and cannot be applied in the practical application of wastewater treatment plants. Mirbagheri et al. [19] selected six of the process variables, including chemical oxygen demand (COD), concentration of sludge, and SRT, as the auxiliary variables, and the membrane flux as the output variable to establish the radial basis function (RBF) model. The RBF-based soft sensing model can successfully predict the permeability of the membrane. However, because the auxiliary variables are mainly selected by experience, the selected auxiliary variables may be redundant, and the prediction accuracy of the model was still unable to meet the actual requirements.

In the aspect of optimizing the structure of ANN, an optimized BPNN-based prediction model was established to predict the ceramic membrane fouling of aqueous extract of Chinese medicine, in which the optimal number of neurons in the hidden layer, the connecting weights and the thresholds of the network model were improved by using an optimization algorithm [38]. In the experimental process, 207 groups of Chinese medicine aqueous extract data were extracted for network training and prediction. Comparatively, the performance of the improved BPNN model is more stable than the traditional BPNN, the RBF network and the multiple regression analysis. The success rate in reaching the preset goal in 20 random running experiments was as high as 95%, which can be adapted to the multi-dimensional and nonlinear data collected in the process of ceramic membrane purification of Chinese medicine aqueous extract. It can be used to predict the degree of membrane fouling stably and accurately, which provides an effective method for the prediction and prevention of fouling of ceramic membrane used to filter the aqueous extracts of Chinese medicine. However, the effectiveness of this prediction model for different raw water has not been mentioned, and needs to be further verified.

To effectively and accurately predict membrane flux of MBR, an improved extreme learning machine (ELM) prediction model was proposed by Yang and Li [39]. Aiming at overcoming the shortcomings of BPNN, such as easily falling into local minimum and poor generalization performance, ELM can effectively overcome these and can obtain good generalization performance at extremely high speed. Since the input weights and hidden layer thresholds are given randomly, ELM usually needs more hidden layer nodes to achieve the desired accuracy [40]. Particle swarm optimization (PSO) was used to optimize the weights and thresholds of ELM to establish the prediction model of PSO-ELM. The principal components extracted by a principal component analysis (PCA) dimension reduction algorithm was used as the input of the model, and the membrane flux was used as the output of the model. The results show that the model has better generalization ability and higher prediction accuracy for MBR membrane flux prediction. In their study, three variables are selected as inputs, including the influent pressure, effluent pressure and corrected flow rate, while other important factors such as mixed liquid suspended solids (MLSS) and temperature were not considered. Therefore, the practicability of the model was studied by Tang et al. [41]. They first used the PCA method to determine the main factors which had significant influence on membrane fouling, including MLSS, operating pressure, and temperature. Then, a prediction model based on the RBF neural network was established. To some extent, high prediction accuracy depends on the parameter selection of network model [42]. Hence, a genetic algorithm (GA) was introduced to the model to optimize the parameters and achieve good global performance. Finally, the prediction results were compared with the measured data. The results illustrated that the convergence speed and prediction accuracy of the membrane fouling simulator based on GA-RBF was better than the RBF network, and the expected goal was achieved. The whole experimental process had a certain theoretical value and practical significance, which would play a positive role in guiding the practical application of MBR.

Li et al. introduced an improved PSO algorithm to the fuzzy RBF neural network to give it a strong nonlinear approximation ability, a self-learning ability, an adaptive ability, and a transient/steady-state performance [43]. Tao et al. employed an improved PSO algorithm to train the fuzzy RBF neural network [44]. The initial weights and thresholds of the fuzzy RBF neural network are obtained by using the improved PSO algorithm, and then the final weights and thresholds are obtained by a secondary optimization. Their results showed that the PSO-based fuzzy RBF neural network is feasible for MBR membrane fouling prediction, which shortens the response time, has a small steady-state error, and can better fit with the actual membrane flux and better predict this.

Based on the RBF neural network, Han et al. [20] proposed a soft sensing method based on a recurrent radial basis function neural network (RRBFNN). Firstly, based on the actual operating data of a wastewater treatment process, the partial least squares (PLS) method was applied to screen out the process variables related to membrane permeability. Secondly, the soft sensing model of membrane permeability was established based on RRBFNN, and the parameters of RRBFNN were adjusted by fast gradient descent algorithm to ensure the accuracy of the soft sensing model. Finally, the designed soft sensing model of membrane permeability was applied to the actual sewage treatment process, and the model was verified by the measured data of wastewater treatment plant. Results showed that the soft sensing model can accurately predict membrane permeability and has a good prediction accuracy. The parameters to be adjusted are output weight $w(t)$, feedback weight $u(t)$, the center of hidden layer neuron $c(t)$ and the width of hidden layer neuron $\sigma(t)$. The updated formula is:

$$\begin{cases} c(t+1) = c(t) - \eta_1 \varphi_c(t) w(t) e(t) \\ \sigma(t+1) = \sigma(t) - \eta_2 \varphi_\sigma(t) w(t) e(t) \\ u(t+1) = u(t) - \eta_3 \varphi_u(t) w(t) e(t) \\ w(t+1) = w(t) - \eta_4 \varphi_w(t) e(t) \end{cases} \quad (1)$$

where $\eta_1, \eta_2, \eta_3, \eta_4$ are the learning rates of $c(t), \sigma(t), u(t)$ and $w(t)$ respectively; and $e(t)$ is the difference between the actual output and the calculated output of the neural network at time t. When the neural network is initialized, the network parameters are set to random values. After the network initialization, the parameters of the neural network are judged and adjusted according to the error value. The calculation formulae of $\varphi_c(t), \varphi_\sigma(t), \varphi_u(t)$, and $\varphi_w(t)$ are:

$$\begin{cases} \varphi_c(t) = \partial \theta(t)/\partial c(t) \\ \varphi_\sigma(t) = \partial \theta(t)/\partial \sigma(t) \\ \varphi_u(t) = \partial \theta(t)/\partial u(t) \\ \varphi_w(t) = \theta(t) - \varphi_c(t) c(t) - \varphi_\sigma(t) \sigma(t) - \varphi_u(t) u(t) \end{cases} \quad (2)$$

The advantage of the fast gradient descent algorithm is that when there are many data to be processed, the calculation speed of neural network will not reduce, and this can ensure the convergence of the algorithm, which is suitable for the use of practical sewage treatment problems. The network output error is:

$$(t) = y(t) - y_d(t) \qquad (3)$$

where $y_d(t)$ is the measured data at time t.

The mechanistic mathematical model [45–48] and computational fluid dynamics simulation [49,50] show excellent performance in a large number of simulation studies of transmembrane pressure difference and flux. However, the complexity of membrane fouling limits the development of these models. To obtain practical kinetic equations, it is necessary to simplify the operating conditions and wastewater characteristics. On this basis, an ANN model based on a recurrent network is proposed to predict the evolution of hydraulic resistance, sediment thickness, and infiltration flux respectively, with TMP and other inflow characteristics as inputs [51,52]. However, in these neural network models, only one set of initial input values is used to predict the whole output set. Geissler et al. [53] used an ANN model to predict the changing trend of membrane flux in MBR. The input parameters are TMP, rate of change of TMP, length of filtration cycle and backwash cycle length, SRT, total suspended solids, aerobic zone temperature, and oxygen decay rate, while the output is membrane flux. Other important parameters such as COD, dissolved oxygen (DO) and MLSS were not considered in their research. In fact, due to a large number of hidden neurons, the predicted value of membrane flux may be over-fitted.

A literature review found that research on predicting membrane fouling in forward osmosis membrane filtration process by the ANN is rarely exercised. Jawada et al. [54] constructed a multilayer neural network model to predict the permeation flux of forward osmosis. The model studied the influence of the number of neurons and hidden layers on the performance of the neural network, which is helpful in optimizing the development of the network structure. The coefficient of determination (R^2) value of the optimized network is as high as 97.3%, which shows the effectiveness of the model in predicting the target yield. Moreover, the effectiveness and generalized prediction ability of the model are verified by some untrained data. The performance of the neural network model is compared with the mass transfer model and multiple linear regression (MLR) model. The results show that the proposed model is superior in the prediction of fouling in the forward osmosis membrane.

The ANN based prediction model can achieve better accuracy by avoiding problems such as the determination of network topology and local extremum value [55]. Based on statistical learning theory and structural risk principle, the support vector machine (SVM) maps the problem to be solved into a high-dimensional space and transforms it into a quadratic optimization problem, which solves the local extremum problem of the neural network. Compared with the traditional SVM, the least squares support vector machine (LSSVM) transforms inequality constraints into equality constraints and transforms the solving process into solving a group of equations, which significantly speeds up, relatively [56]. However, the problem of parameter selection of LSSVM has seriously hindered its development. Nie et al. [57] used the GA algorithm for parameter selection of LSSVM and proposed an MBR membrane flux prediction algorithm based on a GA based LSSVM. To select the parameters of LSSVM accurately, GA is used to optimize the penalty factor and kernel function parameters of the LSSVM model. Meanwhile, PCA was used to determine the main factors that affect MBR membrane flux. The important factors were extracted as the input layer of LSSVM, and the membrane flux was taken as the output. The GA-LSSVM prediction model had a high prediction accuracy for membrane flux.

Zhou et al. [58] established a BPNN model for single-step prediction based on data interpolation and multi-step memory, with the influent water quality concentration of BOD, total nitrogen (TN), total phosphorus (TP), sludge concentration, and influent flow as the auxiliary variables. This model was used to predict the BOD, TN, and TP in the effluent. However, it was only capable of single-step prediction of the parameters mentioned above.

During wastewater treatment, the membrane flux is often used to evaluate the fouling of membrane, and a multi-step prediction of the permeability rate can bring more economic benefits. Therefore, Han et al. [21] proposed a multi-step prediction method of MBR permeability rate based on FNN. Moreover, a soft sensor model was established using the improved Levenberg Marquardt (L-M) algorithm to adjust and optimize the center, width, output weight, and other parameters in a fuzzy neural network (FNN) [59]. In this study, auxiliary variables are taken as the inputs of FNN, while permeability was considered as the output. A multi-step prediction method based on time difference (TD) is proposed to achieve high accuracy. With this method, error accumulation when predicting auxiliary variables is alleviated, and the prediction accuracy is improved. Multi-step prediction is a strategy to achieve value predictions by iterating the single-step prediction procedure, which aims to obtain the output at time step $(t+1)$ time from the variables at time step t:

$$\hat{x}_{t+1} = f(x_t(1), x_t(2), \ldots, x_t(n)) \tag{4}$$

where, $\hat{x}_{t+1}$ is the predicted output at time $t+1$, and $x_i(1), x_i(2), \ldots, \hat{x}_{t+1}$ are n observations at time t.

Indirect multi-step prediction integrates the output of single-step prediction into the input for the subsequent calculation:

$$\hat{x}_{t+m} = f(x_t(1), x_t(2), \ldots, x_t(n), \hat{x}_{t+1}, \ldots, \hat{x}_{t+m-1}) \tag{5}$$

The TD method combines dynamic programming and Monte Carlo sampling algorithms. It is capable of updating the function values in a single and rapid step. The updating formula of the TD method is as follows.

$$Z(t+1) = Z(t) + \alpha(G(t) - Z(t)) \tag{6}$$

where Z is the auxiliary variable, α is the deviation, and $G(t)$ is the updated return value.

Altunkaynak et al. [60] established a soft sensing model of water permeability based on a fuzzy neural network (FNN). In the model, initial flux, inlet shear rate, instantaneous pressure, and filtration time were selected as auxiliary variables based on the mechanism analysis. The model was based on fuzzy mathematics and produces uncertain results.

The osmotic membrane bioreactor (OMBR) is a new technique for wastewater treatment, but its greatest challenge is membrane fouling. The membrane flux of OMBR was simulated and predicted by the adaptive network-based fuzzy inference system (ANFIS) and ANN models based on the adaptive network [61]. MLSS, EC, and DO were used as the model inputs. The ANFIS model and the ANN models were used to predict the parameters of thin-film composite (TFC) and cellulose triacetate (CTA) membranes, and the models' R^2 were 0.9755 and 0.9404 for TFC and 0.9861 and 0.9817 for CTA, respectively. The root means square errors of TFC (0.2527) and CTA (0.1230) in the ANFIS model were lower than those in the ANN model, which were 0.4049 and 0.1449, respectively. Besides, RMSE, SSE, Adj-R^2, and R^2 were used to compare the two established models, and the results showed that the ANFIS model possessed better prediction ability than the ANN model [62].

To solve the problem of membrane fouling, a membrane cleaning decision model was established with membrane recovery as the decision indices. In this model, Bandelet transform was combined with a neural network, creating a Bandelet neural network to predict membrane flux and recovery. However, the Bandelet neural network still has some limitations, such as its high sensitivity to initial weights, its tendency to fall into the local optimal solution during optimization, and the overfitting problem [63,64].

Zhao et al. [65] combined Bandelet transform with a neural network and designed a Bandelet neural network to predict membrane flux and recovery rate so that reasonable decisions on membrane cleaning could be made. Specifically, Bandelet function and scale function were used as the activation functions of the hidden layer and the output layer, respectively. Besides, the improved Bat algorithm [66] was integrated in the Bandelet neural network to improve the optimization outcome. The model is advantageous in its prediction accuracy and speed, and the prediction results are better than those produced by other prediction models.

3. Prediction of Membrane Fouling Based on Mathematical Models

In the current prediction models of membrane fouling during coagulation–membrane filtration, the extended Derjaguin-Landau-Verwey-Overbeek (XDLVO) theory has generally been used to calculate the activation energy of smooth interfaces, but the surface morphology of coagulation flocs will have a greater impact on the prediction results than the activation energy [67]. The sine wave sphere model was used to simulate the surface of rough humic acid (HA) flocs, and a strategy combining surface element integration (SEI), XDLVO theory and composite Simpson rule was used to quantitatively simulate the interfacial interaction energy between different rough flocs and the polyvinylidene fluoride (PVDF) membrane. The results obtained from the combination strategy and the traditional XDLVO method were compared, and the measured interaction energy of smooth interface was compared with that obtained from theoretical simulation. The comparison showed that the combination strategy is suitable to simulate the interfacial interaction energy of flocs in a system of coagulation–membrane filtration. Meanwhile, the roughness of flocs will lead to a difference of 1–2 orders of simulated magnitude in the interfacial interaction energy. Besides, the rough flocs fit the membrane fouling trend better than the smooth flocs. In other words, rough flocs' surface morphology can introduce interactions between flocs and membrane interface, facilitating a higher degree of confidence in characterizing the tendency of membrane fouling [68].

In membrane bioreactors, the interfacial interactions determine the adhesion and membrane fouling caused by pollutants. Therefore, it is of great significance to propose an effective method to quantitatively analyze interfacial interactions [69]. The RBF and ANN were used to predict the interfacial interactions between rough film surfaces. The interaction data were quantified by the XDLVO method and used as the training samples for the RBF network. The results showed that, under the same conditions, the RBF neural network only needed about 1/50 of the calculation time compared to the advanced XDLVO method to obtain the predictions, indicating the higher prediction efficiency of the RBF neural network [70]. Meanwhile, the RBF neural network produced reliable results with acceptable accuracy, suggesting its broad application prospects in the study of membrane fouling and interfacial behaviors.

Specific flux is the main index of membrane permeability. It is defined as:

$$J = J_{20} \cdot 1.025^{(T-20)} \tag{7}$$

$$SF = J_{20}/\Delta P \tag{8}$$

where, J_{20} is the membrane flux at 20 °C, L/(m^2 h); T is the temperature in °C; and SF is the specific flux or permeability of the membrane, L/(m^2·h·kPa).

Xu et al. [71] used the specific flux decay rate with cumulative water yield to evaluate membrane life and established a strategy for membrane life prediction from two perspectives, namely the decline of average specific flux in an actual operation and the recovery of membrane permeability after off-line cleaning. In an actual operation of a MBR, when the average specific flux remains too low to meet the requirement of water production for a long time, the membrane is considered to have reached its service life, and membrane replacements will be necessary. The membrane life predicted by this method is denoted as T_{life}-1. In an MBR operation, membrane fouling exacerbates gradually, and off-line cleaning is the main means to remove it and restore membrane permeability. If the specific fluxes before and after off-line cleaning remain the same, then it can be concluded that cleaning failed to restore membrane permeability; in other words, the membrane life has come to an end. The membrane life prediction is denoted as T_{life}-2. By analyzing the long-term operation (>3 years) of three large-scale municipal wastewater MBR treatment projects

(>10,000 m³/d) and the effects of NaOCl off-line cleaning, two methods were proposed to predict the membrane life: one was based on the descending trend of actual average specific flux and the other was based on the recovery of membrane permeability before and after off-line cleaning. The results showed that for a specific MBR project, T_{life}-2 was slightly larger than T_{life}-1, but when the membrane operation time became longer than T_{life}-1, the actual clean water yield could not meet the requirements. Therefore, T_{life}-1 has more practical engineering significance.

Fenu et al. [22] took a large-scale MBR project as the research object and tried to evaluate the membrane life from different perspectives, such as the failure of water production to meet the designated requirements, the failure of membrane permeability restoration, the maximum strength of membrane contact with the cleaning agent, the continuous increase of operation costs, and the continuous decline of mechanical strength. However, the feasibility and practical application value of Fenu's model need further study.

Wang et al. [72] drew a prediction chart for membrane life in a reclaimed MBR water plant (Figure 3). As the efficiency of chemical cleaning becomes lower and lower, the distance between the upper and the lower curves gradually shrinks and eventually intersects at a point at which a larger flux cannot be obtained by increasing the operating pressure, nor can the water permeability be further recovered by chemical cleaning. Therefore, this intersection point can be regarded as the end of the membrane life in a practical sense. The membrane life was first defined via different concepts, and it was predicted from various perspectives such as water yield, water quality, decay of water permeability, cumulative chlorine contact value, and membrane performance. The results showed that permeability decay and membrane performance are feasible parameters to judge membrane life, while the remaining three parameters are not ideal.

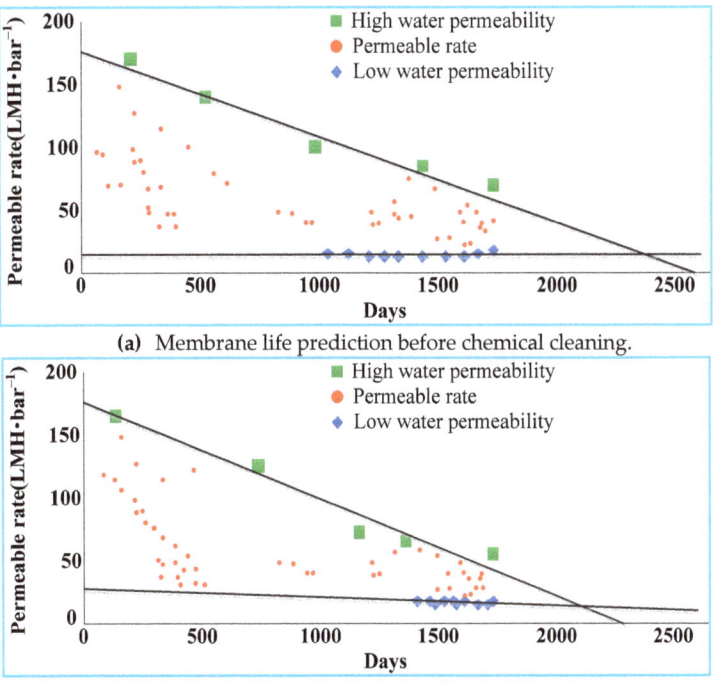

(a) Membrane life prediction before chemical cleaning.

(b) Membrane life prediction after chemical cleaning.

Figure 3. Membrane life prediction.

Various analytical tools were used to systematically study the changes in the removal of chemical oxygen demand (COD), biogas generation, sludge and cake properties, and their correlation with membrane fouling before and after changes to the pH of the feed solution [73]. The results showed that when pH was 8.0, the COD removal rate, gas production rate, and membrane filtration performance of submerged anaerobic membrane bioreactor (SAnMBR) did not show significant alteration. However, when pH was 9.1 or 10.0, the three variables showed a significant change, which lasted for a long time. An increase of pH will lead to the dispersion of sludge flocs and the accumulation of colloids, solutes, and biopolymers in the sludge suspension, resulting in lower membrane performance. Statistical analysis showed that the ratio of protein (PN) to polysaccharide (PS) in extra-cellular polymeric substance (EPS) had a strongly negative correlation with the membrane fouling rate. When pH was 10, under such an alkaline environment, the particles deposited on the membrane surface became smaller and condensed into a more compact cake-like layer, resulting in a higher rate of membrane fouling.

An improved collision adhesion model has been adopted to study the interactions between algae interfaces and hydrodynamic forces without neglecting any hydrodynamic force on the algae. The TOPSIS (technique for order preference by similarity to an ideal solution)—GRA (grey relational analysis) method was also used to evaluate membrane fouling performance, which was studied and predicted by combining the improved collision adhesion model with the back propagation neural network (GA-BP) optimized by a genetic algorithm [74]. The concept of "critical vibration frequency" was put forward, and the energy consumption to separate algae was analyzed, providing theoretical guidance for real-world practices.

Park et al. [24] studied a vertically-oriented hollow fiber membrane module in a pilot-scale bioreactor. The module was equipped with two aerators for simultaneous air injections, one facing downward and another facing upward. Different air jet structures would have significant impacts on the features of membrane fouling, and the experimental results showed that air injection both upward and downward could alleviate membrane fouling effectively. Besides, membrane permeability was revealed to be related to the dynamic information of the variables during membrane treatment. Therefore, further studies are worthwhile to find out whether such correlation can be used to predict membrane fouling.

In a pilot-scale membrane bioreactor (SMBR) for municipal wastewater treatment, the normalized permeability is negatively correlated to the MLSS concentration while positively correlated to the aeration intensity of coarse bubbles. As the concentration-time continued to extend, a small increase in MLSS concentration and mixture viscosity became more common at a certain MLSS concentration, resulting in lower membrane permeability [25].

Some scholars have successfully predicted the variation trend of membrane flux by studying the correlation between membrane flux and several variables in membrane filtration [26,27]. However, membrane fouling is a complex and dynamic process, and the numerous factors affecting membrane permeability and their mutual interactions make it difficult to describe membrane permeability with a simple variable relationship [75]. Martín-Pascual et al. [23] proposed a general mathematical model to estimate membrane permeability and corrected the model parameters with the actual measurements of the process variables. This model is advantageous due to fewer parameters involved and a simple correction process. It has been widely used in the calculation of membrane permeability. However, the parameters included in this model are rather inaccessible for online corrections, and the prediction accuracy is low.

Griffiths et al. [45] developed a mathematical model based on the adhesion probability of particles on the membrane pore wall and the probability of particles falling in a specific pore. In the model, a flat membrane was studied under constant pressure for total flow rate and permeation volume per unit membrane area. The relationship between the total flow rate and the permeation volume per unit membrane area was found to illustrate the main fouling mode at a certain time. Specifically, the fouling modes are standard plugging, in which a certain number of particles adhere to the pore wall and block the pores, and complete plugging, in which larger particles fall on the membrane surface, block the pores completely, and turn into a cake layer. Although this mathematical model can describe these different fouling modes, it cannot predict the TMP or the permeate flux at a given time.

Pimentel et al. [46] proposed a model to reproduce the dynamics of TMP in submerged MBR with infiltration flow rate, aeration rate, solid concentration, and temperature as the inputs. This dynamic model could predict the evolution of TMP for an acceptable period (about 10 days), and the determination coefficient between the predicted TMP and the experimental data was about 0.95. However, the model could not predict TMP evolution in real-time. Charfi et al. [48] studied the anaerobic fluidized bed MBR with granular activated carbon (GAC) and proposed several semi-empirical mathematical models to predict the parameter dynamics, including the dynamics of MLSS concentration or flux. These models could describe the variations of TMP effectively, and its R^2 varied from 95.63% to 99.93%. Besides, they could predict membrane fouling in an anaerobic fluidized bed MBR. Judging from these studies, it could be concluded that as long as the number of input parameters is limited, these mathematical models can effectively predict membrane fouling.

In the field of membrane technologies, it is very difficult to study the internal fluid flow of a membrane module, and one feasible strategy to tackle the problem is to apply computational fluid dynamics (CFD). TMP evolution was simulated in a pilot-scale anaerobic MBR equipped with a flat membrane by ANSYS software. However, membrane optimization by CFD increases the computational cost. Nevertheless, the integration of artificial intelligence (AI) and CFD could facilitate simulation of the membranes and the phase separation process. Babanezhad et al. [76] used the adaptive-network-based fuzzy inference system (ANFIS) model with different parameters to learn about membrane technology. The purpose of the study was to find out how to adjust different parameters in the ANFIS model to improve the prediction accuracy of the AI model on membrane performance. The results indicated that the AI algorithm will obtain a high prediction performance with more input variables. That is, to predict the RUL, the more input variables, the higher the prediction accuracy. When the number of inputs increased to 5, the R-value of the AI predictions increased to 0.99, indicating high accuracy of the AI algorithm in predicting membrane performances.

Recently, Zhang et al. [49] developed a 3D CFD model for open field operation and manipulation (OpenFOAM®) to predict the evolution of permeation flux in MF tubular membranes. The membranes were equipped with baffles as turbulence promoters to reduce particle depositions on the membrane surface. The model produced satisfactory simulations of the MF process. However, CFD has some limitations. The numerical calculation algorithms used in CFD simulation software are based on several selected mathematical models. If a large number of inputs are considered, these models will be too complex to be developed. Besides, the research on CFD prediction of membrane fouling has predominantly focused on membranes with a simple geometric structure. Relevant articles have put forward several basic assumptions to simplify the numerical resolution, which weakens their practical value.

The mathematical models, including the CFD models, are of great significance for the prediction of membrane fouling. However, membrane fouling is a complex phenomenon that is dependent on many factors such as the concentration of extra-cellular polymeric substances and soluble microbial products, which are related to the operation parameters, including biological activity and aeration. Therefore, if fewer parameters are considered and some approximations and assumptions are made to the treatment system, these mathematical models could be calculated with simple equations to simulate the fouling mechanism without complex implementations.

4. Conclusions

MBR plays an important role in wastewater treatment because of its excellent performance. However, the problem of membrane fouling restricts the application of MBR to a great extent. As the technologies for wastewater treatment advance, the requirements for wastewater treatment become higher, and the control of membrane fouling is becoming recognized as the key to overcome the bottleneck of MBR developments. Real-time, fast, and accurate predictions of membrane fouling can not only enhance its control but also reduce the operating costs and improve the efficiency of wastewater treatment. To predict the evolution of membrane resistance, many mathematical methods based on membrane fouling mechanisms have been developed. However, due to the complexity of membrane fouling, these methods remain quite limited, and many assumptions are necessary to simplify them in order to make them feasible for calculations. Artificial neural network was first applied to the predictions of factors related to membrane fouling due to its good modeling capability, and indeed achieved satisfactory results in a very short time. It has also been used to establish an input-output prediction model based on practical membrane fouling data. However, since membrane fouling mechanisms are complex and it is difficult to collect data, the establishment of an ANN-based membrane fouling prediction model still faces many challenges. To solve the problems mentioned above, further research should focus on the following aspects:

(1) Membrane fouling mechanisms in MBRs of different structures and scales should be studied. An important premise of accurate and rapid membrane fouling prediction is the thorough understanding of the underlying mechanisms. Meanwhile, the development of an accurate and real-time online collection system of membrane fouling data can help to build a more comprehensive prediction model with higher prediction accuracy.

(2) Further research should focus on remaining useful life (RUL) prediction of the membrane modules at various failure modes. Most of the current research has focused on the residual life prediction at a single failure mode, ignoring that the failure of the membrane modules is usually caused by the synergistic effect of multiple failure modes. Under certain external impacts, the membranes could suddenly fail to provide normal functions. Therefore, residual life prediction at various failure modes is worthy of further study.

(3) Intelligent feature extraction and remaining useful life prediction should be addressed in future research. An accurate prediction of remaining useful life of membrane components is dependent on the extraction of effective information from the large amount of data obtained from monitoring. However, traditional extraction methods for statistical data and shallow machine learning strategies need to rely on a large number of signal data and expert experience to extract the feature information manually. When processing a large amount of monitoring data from complex engineering equipment, these subjective data extraction methods are seriously limited. Deep learning, such as deep belief network and convolutional neural network, can overcome such problems to some extent, but relevant research is still scarce, suggesting the necessity for further research.

Funding: This research was funded by the Natural Science Foundation of China under grant no. 61863026. Also, it is partially supported by the Industrial Support and Guidance Project for Higher Education of Gansu Province (2019C-05) and the Open Fund Project of Key Laboratory of Gansu Advanced Control for Industrial Process (2019KFJJ03).

Conflicts of Interest: The authors declare no conflict of interest.

References

1. Lares, M.; Ncibi, M.C.; Sillanpaa, M.; Sillanpaa, M. Occurrence, identification and removal of microplastic particles and fibers in conventional activated sludge process and advanced MBR technology. *Water Res.* **2018**, *133*, 236–246. [CrossRef] [PubMed]
2. Krzeminski, P.; Leverette, L.; Malamis, S.; Katsou, E. Membrane bioreactors—A review on recent developments in energy reduction, fouling control, novel configurations, LCA and market prospects. *J. Membr. Sci.* **2017**, *527*, 207–227. [CrossRef]
3. Du, X.J.; Shi, Y.K.; Jegatheesan, V.; Haq, I.U. A review on the mechanism, impacts and control methods of membrane fouling in MBR system. *Membranes* **2020**, *10*, 24. [CrossRef] [PubMed]
4. Alturki, A.A.; Tadkaew, N.; McDonald, J.A.; Khan, S.J.; Price, W.E.; Nghiem, L.D. Combining MBR and NF/RO membrane filtration for the removal of trace organics in indirect potable water reuse applications. *J. Membr. Sci.* **2010**, *365*, 206–215. [CrossRef]
5. Miura, Y.; Watanabe, Y.; Okabe, S. Significance of Chloroflexi in performance of submerged membrane bioreactors (MBR) treating municipal wastewater. *Environ. Sci. Technol.* **2007**, *41*, 87–94. [CrossRef] [PubMed]
6. Shin, C.; McCarty, P.L.; Kim, J.; Bae, J. Pilot-scale temperate-climate treatment of domestic wastewater with a staged anaerobic fluidized membrane bioreactor (SAF-MBR). *Bioresour. Technol.* **2014**, *159*, 95–103. [CrossRef]
7. Abegglen, C.; Joss, A.; McArdell, C.S.; Fink, G.; Schlusener, M.P.; Ternes, T.A.; Siegrist, H. The fate of selected micropollutants in a single-house MBR. *Water Res.* **2009**, *43*, 2036–2046. [CrossRef]
8. Chu, H.P.; Li, X.-Y. Membrane fouling in a membrane bioreactor (MBR): Sludge cake formation and fouling characteristics. *Biotechnol. Bioeng.* **2005**, *90*, 323–331.
9. Zhao, C.Q.; Xu, X.C.; Chen, J.; Wang, G.W.; Yang, F.L. Highly effective antifouling performance of PVDF/graphene oxide composite membrane in membrane bioreactor (MBR) system. *Desalination* **2014**, *340*, 59–66. [CrossRef]
10. Verrecht, B.; Maere, T.; Nopens, I.; Brepols, C.; Judd, S. The cost of a large-scale hollow fibre MBR. *Water Res.* **2010**, *44*, 5274–5283. [CrossRef]
11. Jegatheesan, V.; Pramanik, B.K.; Chen, J.Y.; Navaratna, D.; Chang, C.Y.; Shu, L. Treatment of textile wastewater with membrane bioreactor: A critical review. *Bioresour. Technol.* **2016**, *204*, 202–212. [CrossRef]
12. Dolar, D.; Gros, M.; Rodriguez-Mozaz, S.; Moreno, J.; Comas, J.; Rodriguez-Roda, I.; Barcelo, D. Removal of emerging contaminants from municipal wastewater with an integrated membrane system, MBR-RO. *J. Hazard. Mater.* **2012**, *239*, 64–69. [CrossRef] [PubMed]
13. Du, X.J.; Shi, Y.K.; Jegatheesan, V.; Liyanaarachch, S. A new hybrid RO/FO system and its digital simulation. *Membr. Sci. Tech.* **2020**, *40*, 117–127.
14. Linares, R.V.; Li, Z.; Yangali-Quintanilla, V.; Ghaffour, N.; Amy, G.; Leiknes, T.; Vrouwenvelder, J.S. Life cycle cost of a hybrid forward osmosis low pressure reverse osmosis system for seawater desalination and wastewater recovery. *Water Res.* **2016**, *88*, 225–234. [CrossRef] [PubMed]
15. Schmitt, F.; Banu, R.; Yeom, I.T.; Do, K.U. Development of artificial neural networks to predict membrane fouling in an anoxic-aerobic membrane bioreactor treating domestic wastewater. *Biochem. Eng. J.* **2018**, *133*, 47–58. [CrossRef]
16. Shakil, M.; Elshafei, M.; Habib, M.A.; Maleki, F.A. Soft sensor for nox and o using dynamic neural networks. *Comput. Electr. Eng.* **2009**, *35*, 5578–5864. [CrossRef]
17. Chellam, S. Artificial neural network model for transient crossflow microfiltration of polydispersed suspensions. *J. Membr. Sci.* **2005**, *258*, 35–42. [CrossRef]
18. Al-Zoubi, H.; Hilal, N.; Darwish, N.A.; Mohammad, A.W. Rejection and modelling of sulphate and potassium salts by nanofiltration membranes: Neural network and Spiegler-Kedem model. *Desalination* **2006**, *206*, 42–60. [CrossRef]
19. Mirbagheri, S.A.; Bagheri, M.; Bagheri, Z. Evaluation and prediction of membrane fouling in a submerged membrane bioreactor with simultaneous upward and downward aeration using artificial neural network-genetic algorithm. *Process. Saf. Environ. Prot.* **2015**, *96*, 111–124. [CrossRef]
20. Han, H.G.; Zhang, S.; Qiao, J.F. Soft-sensor Method for Permeability of the Membrane Bio-Reactor Based on Recurrent Radial Basis Function Neural Network. *J. Beijing Univ. Technol.* **2017**, *43*, 1168–1174.
21. Han, H.G.; Zhang, Q.; Qiao, J.F. Multi-step prediction of permeability of the membrane bio-reactor based on fuzzy neural network. In Proceedings of the 38th Chinese Control Conference, Guangzhou, China, 27–30 July 2019.
22. Fenu, A.; Wilde, W.D.; Gaertner, M.; Weemaes, M.; Gueldre, G.D. Elaborating the membrane life concept in a full scale hollow-fibers MBR. *J. Membr. Sci.* **2012**, *421–422*, 349–354.
23. Martín-Pascual, J.; Leyva-Díaz, J.C.; López-López, C.; Munñio, M.M.; Hontoria, E.; Poyatos, J.M. Effects of temperature on the permeability and critical flux of the membrane in a moving bed membrane bioreactor. *Desalin. Water Treat.* **2015**, *53*, 3439–3448. [CrossRef]
24. Park, H.D.; Lee, Y.H.; Kim, H.B. Reduction of membrane fouling by simultaneous upward and downward air sparging in a pilot-scale submerged membrane bioreactor treating municipal wastewater. *Desalination* **2010**, *251*, 75–82. [CrossRef]
25. Trussell, R.S.; Merlo, R.P.; Hermanowicz, S.W. Influence of mixed liquor properties and aeration intensity on membrane fouling in a submerged membrane bioreactor at high mixed liquor suspended solids concentrations. *Water Res.* **2007**, *41*, 947–958. [CrossRef]
26. Hwang, B.K.; Lee, W.N.; Park, P.K. Effect of membrane fouling reducer on cake structure and membrane permeability in membrane bioreactor. *J. Membr. Sci.* **2007**, *288*, 149–156. [CrossRef]
27. Luna, H.J.; Baeta, B.E.L.; Aquino, S.F. EPS and SMP dynamics at different heights of a submerged anaerobic membrane bioreactor (SAMBR). *Process. Biochem.* **2014**, *49*, 2241–2248. [CrossRef]

28. Jiao, L.C.; Yang, S.Y.; Liu, F.; Wang, S.G.; Feng, Z.X. Seventy years beyond neural networks: Retrospect and prospect. *Chin. J. Comput.* **2016**, *39*, 1697–1716.
29. Yao, Y.D.; He, J.J.; Li, Y.L.; Xie, D.Y.; Li, Y. ET0 simulation of self-constructed improved whale optimized BP neural network. *J. Jilin Univ.* **2021**, 1–10. [CrossRef]
30. Fan, Z.; Tian, R.Z.; Lin, L.; Han, Y.Z.; Guo, Y.; Dou, L.L.; Jing, G.H.; Agi, D.T. Desulfurization optimization of reforming catalytic dry gas using radial basis artificial neural network based on PSO algorithm. *Chem. Ind. Eng. Prog.* **2021**, 1–15. [CrossRef]
31. Wang, J.; Cao, J.X.; Zhou, X. Reservoir Porosity Prediction Based on Deep Bidirectional Recurrent Neural Network. *Prog. Geophys.* **2021**, 1–10. Available online: http://kns.cnki.net/kcms/detail/11.2982.P.20210208.1016.052.html (accessed on 20 May 2021).
32. Liu, G.G.; Pei, L.Y.; Yang, Y.M.; Li, S.N. Compactness prediction of airport soil field based on artificial neural network. *J. Shenzhen Univ. Sci. Eng.* **2021**, *38*, 54–60.
33. Dornier, M.; Trystram, G.; Lebert, A. Dynamic modeling of crossflow microfiltration using neural networks. *J. Membr. Sci.* **1995**, *98*, 263–273. [CrossRef]
34. Shetty, G.R.; Chellam, S. Predicting membrane fouling during municipal drinking water nanofiltration using artificial neural networks. *J. Membr. Sci.* **2003**, *217*, 69–86. [CrossRef]
35. Delgrange, N.; Cabassud, C.; Cabassud, M.; Durand-Bourlier, L.; Lainé, J.M. Neural networks for prediction of ultrafiltration transmembrane pressure—Application to drinking water production. *J. Membr. Sci.* **1998**, *11*, 11–123. [CrossRef]
36. Li, W.W.; Li, C.Q.; Wang, T. Application of machine learning algorithms in MBR simulation under big data platform. *Water Pract. Technol.* **2020**, *15*, 1–2. [CrossRef]
37. Montserrat, D.A.; Nataa, A.B.; Sara, G.A. Comparison of a deterministic and a data driven model to describe mbr fouling. *Biochem. Eng. J.* **2015**, *260*, 300–308.
38. Dou, P.W.; Wang, Z.; She, K.K.; Fan, W.L. Study on Forecasting Ceramic Membrane Fouling in TCM Extracts Based on Improved BP Neural Network. *Chin. J. Inf. TCM* **2017**, *24*, 92–96.
39. Yang, X.X.; Li, C.Q. Research of MBR simulation predictions based on improved extreme learning machine. *Comput. Eng. Softw.* **2016**, *37*, 17–20.
40. Wang, H.; Wang, Y.; Ji, Z.C. Simulation of Wind Power Prediction Based on Improved ELM. *J. Syst. Simul.* **2018**, *30*, 4437–4447.
41. Tang, J.; Li, C.Q. Research of RBF neural network based on genetic algorithm optimization in MBR membrane pollution simulation. *Comput. Eng. Softw.* **2016**, *19*, 11–13.
42. Jin, Y.; Liu, S.J.; Zhang, J.G. Fatigue reliability of high speed bearing based on genetic algorithm optimized artificial neural network. *J. Aerosp. Power* **2018**, *33*, 2748–2755.
43. Li, J.J.; Li, X.F.; Pian, J.X. Temperature control of annealing furnaces based on improved PSO and fuzzy RBF neural network. *J. Nanjing Univ. Sci. Technol.* **2014**, *38*, 337–341.
44. Tao, Y.X.; Li, C.Q.; Su, H. Prediction of MBR membrane pollution based on improved PSO and fuzzy RBF neural network. *Comput. Eng. Softw.* **2018**, *39*, 52–56.
45. Griffiths, I.M.; Kumar, A.; Stewart, P.S. A combined network model for membrane fouling. *J. Colloid Interface Sci.* **2014**, *432*, 10–18. [CrossRef]
46. Pimentel, G.A.; Dalmau, M.; Vargas, A.; Comas, J.; Rodriguez-Roda, I.; Rapaport, A.; Vande Wouwer, A. Validation of a simple fouling model for a submerged membrane bioreactor, IFAC-Pap. *Online* **2015**, *48*, 737–742.
47. Zuthi, M.F.R.; Guo, W.; Ngo, H.H.; Nghiem, D.L.; Hai, F.I.; Xia, S.; Li, J.; Liu, Y. New and practical mathematical model of membrane fouling in an aerobic submerged membrane bioreactor. *Bioresour. Technol.* **2017**, *238*, 86–94. [CrossRef]
48. Charfi, A.; Aslam, M.; Lesage, G.; Heran, M.; Kim, J. Macroscopic approach to develop fouling model under GAC fluidization in anaerobic fluidized bed membrane bioreactor. *J. Ind. Eng. Chem.* **2017**, *49*, 219–229. [CrossRef]
49. Zhang, W.; Ruan, X.; Ma, Y.; Jiang, X.; Zheng, W.; Liu, Y.; He, G. Modeling and simulation of mitigating membrane fouling under a baffle-filled turbulent flow with permeate boundary. *Sep. Purif. Technol.* **2017**, *179*, 13–24. [CrossRef]
50. Boyle-Gotla, A.; Jensen, P.D.; Yap, S.D.; Pidou, M.; Wang, Y.; Batstone, D.J. Dynamic multidimensional modelling of submerged membrane bioreactor fouling. *J. Membr. Sci.* **2014**, *467*, 153–161. [CrossRef]
51. Piron, E.; Latrille, E.; René, F. Application of artificial neural networks for crossflow microfiltration modelling: Black-box and semi-physical approaches. *Comput. Chem. Eng.* **1997**, *21*, 1021–1030. [CrossRef]
52. Hamachi, M.; Cabassud, M.; Davin, A. Mietton Peuchot, M. Dynamic modelling of crossflow microfiltration of bentonite suspension using recurrent neural networks. *Chem. Eng. Process. Process. Intensif.* **1999**, *38*, 203–210. [CrossRef]
53. Geissler, S.; Wintgens, T.; Melin, T.; Vossenkaul, K.; Kullmann, C. Modelling approaches for filtration processes with novel submerged capillary modules in membrane bioreactors for wastewater treatment. *Desalination* **2005**, *178*, 125–134. [CrossRef]
54. Jawada, J.; Hawarib, A.H.; Zaidia, S. Modeling of forward osmosis process using artificial neural networks (ANN) to predict the permeate flux. *Desalination* **2020**, *484*, 114427. [CrossRef]
55. Wang, Z.; Wang, Y.F. Creep rupture life estimation of P91 steel pipes through artificial neural network based hardness prediction. *J. Chin. Soc. Power Eng.* **2020**, *40*, 936–939.
56. Yu, W.L.; Wang, Y.; Ji, Z.C. Modeling and process parameters optimization of GlcN fermentation process based on improved MVO-LSSVM. *J. Syst. Simul.* **2020**, *32*, 1–8.
57. Nie, J.Y.; Li, C.Q.; Li, W.W.; Wang, T. Research on the least squares support vector machine optimized by genetic algorithm in the simulation MBR prediction. *Comput. Eng. Softw.* **2015**, *36*, 40–44.

58. Zhou, Y.; Chang, F.J.; Chang, L.C.; Kao, I.F.; Wang, Y.S. Explore a deep learning multi- output neural network for regional multi-step-ahead air quality forecasts. *J. Clean. Prod.* **2019**, *209*, 134–145. [CrossRef]
59. Wilamowski, B.M.; Yu, H. Improved computation for Levenberg-Marquardt training. *IEEE Trans. Neural Netw.* **2010**, *21*, 930–937. [CrossRef] [PubMed]
60. Altunkaynak, A.; Chellam, S. Prediction of specific permeate flux during crossflow microfiltration of polydispersed colloidal suspensions by fuzzy logic models. *Desalination* **2010**, *253*, 188–194. [CrossRef]
61. Hosseinzadeh, A.; Zhou, J.L.; Altaee, A.; Baziar, M.; Li, X.W. Modeling water flux in osmotic membrane bioreactor by adaptive networkbased fuzzy inference system and artificial neural network. *Bioresour. Technol.* **2020**, *310*, 123391. [CrossRef] [PubMed]
62. Hosseinzadeh, A.; Baziar, M.; Alidadi, H.; Zhou, J.L.; Altaee, A.; Najafpoor, A.A.; Jafarpour, S. Application of artificial neural network and multiple linear regression in modeling nutrient recovery in vermicompost under different conditions. *Bioresour. Technol.* **2020**, *303*, 122926. [CrossRef]
63. Alami, N.; Meknassi, M.; En-nahnahi, N. Enhancing unsupervised neural networks based text summarization with word embedding and ensemble learning. *Expert Syst. Appl.* **2019**, *123*, 195–211. [CrossRef]
64. Sudakov, O.; Burnaev, E.; Koroteev, D. Driving digital rock towards machine learning: Predicting permeability with gradient boosting and deep neural networks. *Comput. Geosci.* **2019**, *127*, 91–98. [CrossRef]
65. Zhao, B.; Chen, H.; Gao, D.K.; Xu, L.Z.; Zhang, Y.Y. Cleaning decision model of MBR membrane based on Bandelet neural network optimized by improved Bat algorithm. *Appl. Soft Comput. J.* **2020**, *91*, 106211. [CrossRef]
66. Zhang, H.F.; Fan, X. Research progress in membrane fouling in membrane bioreactor based on XDLVO approach. *Chemistry* **2016**, *79*, 604–609.
67. Yang, X.S.; Gandomi, A.H. Bat algorithm: A novel approach for global engineering optimization. *Eng. Comput.* **2012**, *29*, 464–483. [CrossRef]
68. Fang, C.; Yang, S.; Wu, Y.; Zhang, H.W.; Wang, J.; Wang, L.T.; Hao, S.Z. Effect of floc surface morphology on membrane pollution prediction. *Ciesc J.* **2020**, *71*, 715–723.
69. Chen, J.; Lin, H.; Shen, L.; He, Y.; Zhang, M.; Liao, B.Q. Realization of quantifying interfacial interactions between a randomly rough membrane surface and a foulant particle. *Bioresour. Technol.* **2017**, *226*, 220–228. [CrossRef] [PubMed]
70. Zhao, Z.T.; Lou, Y.; Chen, Y.F.; Lin, H.J.; Li, R.J.; Yu, G.Y. Prediction of interfacial interactions related with membrane fouling in a membrane bioreactor based on radial basis function artificial neural network (ANN). *Bioresour. Technol.* **2019**, *282*, 262–268. [CrossRef] [PubMed]
71. Xu, Y.; Lei, T.; Sun, J.Y.; Xia, J.L.; Xue, T.; Yu, K.C.; Guan, J.; Chen, Y.L.; Huang, X. Life-time assessment of membrane in large-scale MBR plants for municipal wastewater treatment. *China Water Wastewater* **2015**, *31*, 34–39.
72. Wang, X.S.; Gao, J.H.; Ai, B.; Fu, Q.; Chang, J.; Gan, Y.P. Evaluation of membrane life span in a full scale MBR process for reclaimed water treatment plant. *Water Purif. Technol.* **2014**, *33*, 24–27.
73. Jane Gao, W.J.; Lin, H.J.; Leung, K.T.; Liao, B.Q. Influence of elevated pH shocks on the performance of a submerged anaerobic membrane bioreactor. *Process. Biochem.* **2010**, *45*, 1279–1287.
74. Jiang, S.H.; Xiao, S.Z.; Chu, H.Q.; Zhao, F.C.; Yu, Z.J.; Zhou, X.F.; Zhang, Y.L. Intelligent mitigation of fouling by means of membrane vibration for algae separation: Dynamics model, comprehensive evaluation, and critical vibration frequency. *Water Res.* **2020**, *182*, 115972. [CrossRef] [PubMed]
75. Alkmim, A.R.; Costa, P.R.; Amaral, M.C.S. The application of filterability as a parameter to evaluate the biological sludge quality in an MBR treating refinery effluent. *Desalin. Water Treat.* **2015**, *53*, 1440–1449. [CrossRef]
76. Babanezhad, M.; Masoumian, A.; Nakhjiri, A.; Marjani, A.; Shirazian, S. Infuence of number of membership functions on prediction of membrane systems using adaptive network based fuzzy inference system (ANFIS). *Sci. Rep.* **2020**, *10*, 16110. [CrossRef] [PubMed]

MDPI
St. Alban-Anlage 66
4052 Basel
Switzerland
Tel. +41 61 683 77 34
Fax +41 61 302 89 18
www.mdpi.com

Membranes Editorial Office
E-mail: membranes@mdpi.com
www.mdpi.com/journal/membranes

www.ingramcontent.com/pod-product-compliance
Lightning Source LLC
LaVergne TN
LVHW070604100526
838202LV00012B/558